CIM
CAD
CAM

Design
Testing

CNC - Mills

Q- control

DNC   Damage
Monitor Nuc Pwr Plnts 301-306

Computer-Controlled
Industrial Machines.
Processes. and Robots

Computer-Controlled Industrial Machines, Processes, and Robots

James A. Gupton, Jr.

PRENTICE-HALL, Englewood Cliffs, New Jersey 07632

Library of Congress Cataloging-in-Publication Data

Gupton, James A.
  Computer-controlled industrial machines, processes,
and robots.

  Bibliography: p.
  Includes index.
  1. Manufacturing processes—Data processing.
I. Title.
TS183.G87 1986   670.42'7   85-28255
  ISBN 0-13-165267-2

Appendix C reproduced courtesy of Fairchild Camera Instrument, Inc.
Appendix D reproduced courtesy of Airpax Corp.
Appendix E reproduced courtesy of Winifred M. Berg, Inc.

Editorial/production supervision
  and interior design: Jane Zalenski
Cover design: Wanda Lubelska
Manufacturing buyer: Gordon Osbourne

Printed in the United States of America

10 9 8 7 6 5 4 3 2 1

ISBN 0-13-165267-2 01

Prentice-Hall International, Inc., *London*
Prentice-Hall of Australia Pty. Limited, *Sydney*
Editora Prentice-Hall do Brasil, Ltda., *Rio de Janeiro*
Prentice-Hall Canada Inc., *Toronto*
Prentice-Hall Hispanoamericana, S.A., *Mexico*
Prentice-Hall of India Private Limited, *New Delhi*
Prentice-Hall of Japan, Inc., *Tokyo*
Prentice-Hall of Southeast Asia Pte. Ltd., *Singapore*
Whitehall Books Limited, *Wellington, New Zealand*

To my wife Edith,
for her patience and understanding
during the months of research and writing
required to complete this book.

Also to my two sons, Jim and Bob,
for their enthusiasm and encouragement
of my work as an author. Thank you, sons.

Contents

PREFACE, xi

ACKNOWLEDGMENTS, xv

1 GENESIS OF AN INDUSTRIAL PROCESS, 1

The MicroMonitor II Process Control System, 3
The MicroMonitor Computer, 7
The MicroMonitor II Functions, 8
The MicroMonitor Control Panels, 9
INFOR/TEX, 12
The Program Control Operation, 17
MicroMonitor Process Control Analysis, 17
Computer-Controlled Machine Shop, 19
Chapter Review Questions, 23
Glossary, 24

2 COMPUTER CONTROL IN A HOSTILE ENVIRONMENT, 26

Contaminants Hazardous to Computers, 26
Peripheral Data Links in a Hostile Environment, 29
Shielded Data Links and Methods of Selection, 30
Data Transmission via Beams of Light, 40
Introduction to Fiber Optics Technology, 41
Chapter Review Questions, 48
Glossary, 49

3 COMPUTERS AND PERIPHERALS, 52

Computer Interface Limitations, 54
Computer Parallel and Serial Interfacing, 63
The Basic Controller BC2, 68
Zilog Industrial BASIC Language, 75
Chapter Review Questions, 78
Glossary, 79

4 PRINCIPLES OF DATA ACQUISITION AND CONVERSION, 81

Quantizing Theory, 83
Sampling Theory, 86
Amplifiers and Filters, 89
Settling Time, 92
Digital Coding, 94
D/A Converters, 96
A/D Converters, 100
Analog Multiplexers, 106
Sample and Hold Circuits, 107
A/D and D/A Converter Adjustments, 108
Chapter Review Questions, 110
Glossary, 111

5 CMOS Technology, 116

CMOS Circuit Performance, 118
The P2CMOS Process, 120
Solving the CMOS Latch-Up Problem, 121
How Speed Has Improved, 122
Specifications for the 54/75HC Logic Family, 123
Comparing Output Current Sink and Source, 124
CMOS and Bipolar Design, 125
CMOS Data Converters Simplify System Design, 125
Chapter Review Questions, 127
Glossary, 128

6 CONTROL DEVICES, INTERFACES, AND MACHINE COMPUTERS, 130

Controllable External Devices, 132
Basic Operation Requirements, 132
Control and Measuring Devices, 135
Control Transducers, 144
Control and Measuring Device Computer Interfaces, 147
Machine Site Computer Installations, 151
Chapter Review Questions, 154
Glossary, 154

**7 ULTRAPRECISION-CONTROLLED DEVICES
AND INTELLIGENT SYSTEMS,** 156

Control by Light and Computer Vision, 161
Machine and Computer Vision, 162
Artificial Intelligence, 175
Chapter Review Questions, 177
Glossary, 178

8 INDUSTRIAL ROBOTS, 182

The Industrial Robot, 183
Industrial Robots on the Production Line, 191
Cincinnati Milacron Industrial Robots, 195
Chapter Review Questions, 203
Glossary, 203

9 FIBER OPTIC TRANSMITTERS-RECEIVERS AND DATA LINKS, 205

Data Transmission and Reception via Optical Frequencies, 206
Fiber Optic Transceivers and Transmitters, 207
Individual Transmitter and Receiver Systems, 211
Computer Interfacing to Fiber Optic Transmitters and Receivers, 217
Chapter Review Questions, 220
Glossary, 221

**10 MODEL ROBOTS AS TEACHING AIDS
FOR ELECTROMECHANICAL ENGINEERING,** 223

Designing a Teaching-Aid Robot, 229
Project Robots with On-Board Computers, 237
The National Semiconductor Industrial Microcomputer System, 250
Chapter Review Questions, 252
Glossary, 253

11 UNIVERSITY RESEARCH ROBOT PROJECTS, 254

Shakey (1967–1969), 254
The Stanford Cart (1973–1981), 255
SCIMR (1981), 256
The Unimation Rover (1983), 256
Robart I (1980–1983), 257
Robart II (1982–1983), 283
References, 286
Chapter Review Questions, 287
Glossary, 288

12 !MAGINEERING: THE FUTURE APPLICATIONS
    FOR COMPUTER CONTROL AND ROBOTICS, 290

Voice-Entry Terminal, 291
Computerized Wheelchair Design Project, 292
Veterans' Administration's Quadriplegic Wheelchair
  with Manipulator, 295
Nuclear Power Plant Damage Control Robot, 302
Mine Rescue Robot, 307
Shipboard Damage Control Robot, 307
Second-Generation Work Robots, 312
References, 312
Chapter Review Questions, 313
Glossary, 314

Appendix A CMOS Technology, 315

Appendix B Decimal-to-Metric Conversion, 322

Appendix C Fairchild CCD-4001 Machine Vision Camera, 325

Appendix D Stepper Motor Handbook, 330

Appendix E Berg Data Sheet, 339

Appendix F Sources, 343

Answers to Odd-Numbered Review Questions, 348

Index, 354

Preface

This text is designed as an entry-level text to electro-mechanical, pneumatic-mechanical, and hydraulic-mechanical machines subject to control by computers. It is suitable for use in a two-semester or three-semester sequence in the area of computer-controlled machines and robots. It would be helpful to the students if they have completed courses in physics, AC and DC circuit analysis, or digital logic circuits through microprocessors. The text has minimal mathematics emphasis in that the only mathematical sections relate to specific formulae associated with chapter subject material.

Each chapter contains learning aids in the form of acronyms common to the subject under discussion, block diagrams, charts, photographic illustrations, electronic and logic circuit schematic diagrams. Students' vocabularies will be expanded through chapter-end glossaries covering specific subject words and their meanings. Self-review questions with selected odd-numbered answers are included, which relate to key points contained in the chapter. The scope of the text ranges from an industrial computer-controlled textile dyeing process and the machines that carry out the process, to industrial computer-controlled robots. In addition, the subjects of data acquisition and conversion, data transmission links, computer vision, speech recognition, and speech synthesis are presented. The robots that were subjects of University Research Technical Reports or Master's theses are of special interest.

Chapter 1 traces the skill of dyeing fabrics and thread from prehistoric times to the first fully-automated textile dyeing machine in 1957, to the first computer-controlled dyeing system in 1978. The major subject area relates to the MicroMonitor II Process Control System and its operating software that achieved total computer control of the textile dyeing process, the dye machine, and the logistics associated with the total process.

Chapter 2 approaches the problems and hazards encountered in computer control in the hostile environments typical of modern industrial facilities. This chapter presents methods and materials to overcome the computer-control interference encountered in industrial problem areas with shielded cables and optical transmission cables as data links.

Chapter 3 emphasizes limitations of computer interface expansion, the architecture of the microprocessor (MPU) with descriptions of the circuitry under its control, and interfacing the MPU to external peripherals through a Peripheral Interface Adapter (Motorola 6280 PIA). This chapter introduces a type of computer and control programming language developed exclusively for industrial control applications.

Chapter 4 presents one of the most comprehensive discussions of data acquisition and conversion. It was prepared by Datel Corporation for customer instruction in Digital-to-Analog/Analog-to-Digital data acquisition and conversion for control and measurement data input to computers.

Chapter 5 presents the latest microelectronics technology in CMOS microcircuits. It explains via diagrams and text the increase of microcircuit chip density, which resulted in decreased data processing time and improved product reliability. A comparison of a CMOS family of logic circuits is given as well as a related appendix covering the National Semiconductor complete family of CMOS products.

Chapter 6 emphasizes specific computer control and measuring devices related to temperature control. Examples of multimeasurement process temperature control and their associated computer interfaces are given. Other examples given in Chapter 6 point out the differences between poor and good control computer installations at a machine site. These good control installations demonstrate the methods to prevent data pollution provided in the subject material of prior chapters.

Chapter 7 covers computer vision and cameras that produce visual images as data input to the computer for inspection and object identification data. Discussions include pixel arrays and resolution capabilities of CCD video cameras used to provide computer vision. The latter part of the chapter provides an introduction to speech technology fundamentals and methods implementing computer speech recognition and voice synthesis.

Chapter 8 introduces the industrial robots concentrating on Unimation and Cincinnati Milacron computer-controlled industrial robots. Emphasis is placed on industrial robot axial movements, axial movement power sources, manipulator working ranges, machine computers, and teach control. This chapter is well illustrated with examples of industrial robots and the types of work they are able to perform.

Chapter 9 presents an excellent background to the understanding of data transmission by light and the types of optical fibers that serve as the core of fiber optic data links. Explanations are given to attenuation and transmission delay factors related to transmission links one kilometer and longer. Special attention is given to stress the immunity exhibited by optical fibers to all forms of data pollution or interference by comparing optical fiber cables to shielded coaxial cables. In addition, coverage is given to computer interfacing to fiber optic cables through operational descriptions of optical transmitters and receivers, including modulation methods, optical output power, and receiver sensitivity.

Chapter 10 explores the concept of robot projects as training aids in electromechanical engineering courses. A number of design considerations are given with examples of small robots using audio tones for control and infrared energy for obstacle avoidance. To facilitate student designs for project robots, a number of robot structural designs are presented in detail covering robot mobility, body superstructure, manipulators, and end effectors. Sources for robot project components are provided in the sources appendix.

Chapter 11 presents a number of university-research robot projects and robots that were the subject of a master's thesis or university research report from 1967 through 1983. The featured robot was the subject of the Master's thesis by LCDR H.R. Everett at the Naval Postgraduate School in Monterey, California (1982). LCDR Everett's robot, Robart I, was the first multisensor, multicontrol-function sentry robot to be the subject of a Master's thesis. Each step in the problem solving of this robot is covered, as well as the operation of its most important navigational and control sensors. The chapter concludes with the subsequent development of Robart II and the improvements incorporated for the wheelbase, body superstructure, and expanding the control computer by making the SYN-1 a supervisory computer over five dedicated control computers.

Chapter 12 carries computer control into the field of medical robot manipulators (for quadriplegic rehabilitation) and a special voice entry terminal that is adaptable for medical manipulator control. The student is given an opportunity to borrow from previous chapter contents to compile control circuits necessary to produce a prototype voice-controlled quadriplegic wheelchair. This chapter further explores the possibilities of future mobile robots designed to effect repair or rescue when damage occurs in hazardous environments such as commercial mines, nuclear power plants, naval vessels, natural disasters, and transportation accidents.

This book contains more than 200 illustrations designed to supplement text descriptions, operational functions and to present visual understanding of a most complex and technical subject on an entry level.

Acknowledgments

The author wishes to acknowledge the assistance and cooperation of the following people without whom this book might never have been completed.

James J. Pinto, *Action Instruments Co. Inc.*

Steven R. Cohen, *Advanced Fiberoptics Corp.*

L. J. Torok, *Airpax Corp. Cheshire Division*

Ellen L. Walyus, *American Photonics*

Norman Uress, *Applied Color Systems*

Judith M. Bondie, *Applied Intelligent Systems*

Thomas Sikes, *Belden Fiber Optics Division*

Gene Tobey, *Burr-Brown Corporation*

Christopher P. Sheehan, *Chartpak Graphics*

Barry A. Spaeth, *Cincinnati Milacron*

Nancy Morris, *Dilithium Press*

Davis Tinis, *D.J.T. Electronics*

Joseph Shrager, *H&R Corporation*

Steve Weitzer, Editor/Bruce Morgen, Associate Editor, *Electronic Products/Hearst Business Communications*

Sandy Furiosi, *Fairchild Camera and Instrument Corporation*

Gordon Hacker, *Gaston County Dyeing Machine Company*

John E. Marsh, *Haydon Switch and Instrument Company*

Susan N. Jerrems, *Micro Robotics Center, Inc.*

Steven McGinness, *National Semiconductor Corporation*

LCDR H. R. Everett, *Naval Sea Command, United States Navy*

David L. Hudson, *Octek, Inc.*

Peter Nero, *Oriel Corporation*

Carl T. Helmers, Jr., Publisher, *Robotics Age, Inc.*

Lee A. Hart, *Technical Micro Systems, Inc.*

Pat J. Rosato, *Unimation, Inc.*

Werner Greenbaum, Acting Director, *Veterans Administration/Rehabilitation Engineering Center*

George Harvey, *Votan*

1

Genesis of an Industrial Process

The goals of yesterday will be the starting point of tomorrow.
Carlyle

In the age of computers and high technology, it is appropriate to follow the lead of Carlyle's quotation in tracing the origin of one of the world's oldest industrial processes.

At the dawn of our present civilization, the first *Homo sapiens* had three goals for survival: finding a source of food, providing protective clothing, and protecting family and territory. Prehistoric cave drawings suggest that natural vegetable and animal tinctorial materials were used by early man to decorate the walls of caves and perhaps to paint their bodies to present a fearsome appearance to their enemies. Archeological diggings in Thebes have revealed that the Egyptians practiced the dyeing of garments with indigo as early as 3500 B.C. Early historical classics of the Chinese from around 2500 B.C. include a number of references to the dyeing of silk. Greek mythology suggests that dyeing came before the skills of spinning and weaving since Ariadne, the goddest of spinning and weaving, was the daughter of Idon, the dyer of wool.

As civilization spread from the Mediterranean basin into what became modern Europe, the Phoenicians practiced the dyeing of fabrics, which became the basis of extensive trade with other nations. Dyeing was practiced as a trade in Rome during the reign of Numa Pompilus, who encouraged and fostered the skill of dyeing by establishing the Collegium Tinctorum, which became the first school devoted to teaching the art and skills of dyeing.

In its earliest form, the dye process consisted of boiling a garment or fabric in a solution of water containing berries, bark, or nutshells. In prehistoric times, this was probably accomplished by filling a depression in rock

Dyeing in the middle ages.

with water and dropping hot rocks into the depression to heat the water and its tinctorial components. In time, the depressions in rock gave way to baked clay vessels and wooden vats. However, as late as the Middle Ages, the dye vats were still being heated with hot rocks to boil the tinctorial liquors.

Although the art of dyeing has been practiced for more than 4500 years, the natural vegetable, animal, and organic chemicals used in the dye process made little or no progress until the discovery of coal-tar coloring chemicals in the mid-1850s. The new dyes, artificially produced from coal tars, were more brilliant and fade resistant than the natural tinctorial products.

We are prone to be egotistical about our accomplishments since the Middle Ages until we face the reality that after 4500 years, the dyeing of fabrics continued to be a manual process until as late as 1940. Progress had been made in several areas: pumps were used to fill the vats, live steam replaced hot rocks to heat the dye liquors, and temperature gauges could accurately measure the dye liquor temperature. However, each improvement remained basically a manually operated control and nothing had been automated. The first automation of a textile dye machine was developed shortly after the end of World War II when an electric motor–driven timer, developed in 1943 to calibrate the Norton bombsight, was modified by replacing the clock hands with a contoured cam. The timer-rotated cam, contoured to profile the dye process temperature cycle, would automatically control the process temperature rise and fall required. It was not until 1957 that a textile dye machine became fully automated by the use of a combination of electromechanical devices and pneumatic analog controls.

Gaston County Dyeing Machine Company, Stanley, North Carolina, has been one of the leading pioneers in the automation of textile dyeing machines. This is the firm that produced the first fully automated dyeing machine in 1957. In the years that followed, Gaston County continued to apply the latest developments in electronic technology toward a goal of total automation of the textile dyeing machines through computer control. Phase I was reached in 1979 with the completion of their first computer-controlled dyeing machine, the MicroMonitor I Process Control System.

Gaston County's research continued into the 1980s, to reach the highest degree of computer use in textile dyeing with their MicroMonitor II Process Control System. This system is most appropriate to introduce the methods by which the computer can be used to control an industrial machine and the industrial process the machine performs.

THE MICROMONITOR II PROCESS CONTROL SYSTEM

When one contemplates the physical size of the present-day microcomputer, it is inconceivable that it could control a machine or machine process that physically occupies three floor levels for its installation. The MicroMonitor II Process Control System is capable of controlling a typical dyehouse operation, such as the one shown in Figure 1.1, and it can control up to 32 identical installations simultaneously. In addition, it can compile the logistics of production, inventory, and process status of each individual machine under the control of the computer.

To better understand the role of a computer in controlling the textile dyeing machines, we must be made aware of the two basic dye processes: batch and continuous sequence. The batch process takes one lot or load of material and completely processes the lot. Figure 1.2 shows the batch-processing keirs in a yarn package dyehouse. Batch-type dyeing machines range in size from sample equipment, capable of dyeing lots up to 1 lb, to production equipment capable of lots weighing up to 3000 lb. Continuous dyeing and bleaching of material is a process whereby a number of steps are sequentially treated until the process has been completed. The computer automatically fills, drains, and switches the dyestuffs according to the sequential step of the process. Figure 1.3 shows a production-size high-temperature, high-pressure, continuous sequential dyeing machine developed for synthetic, double-knit fabrics in a textile dyehouse.

Figure 1.4 illustrates the numerous electropneumatic valves, pumps, and sensors that enable the computer to control a complex dye machine in-

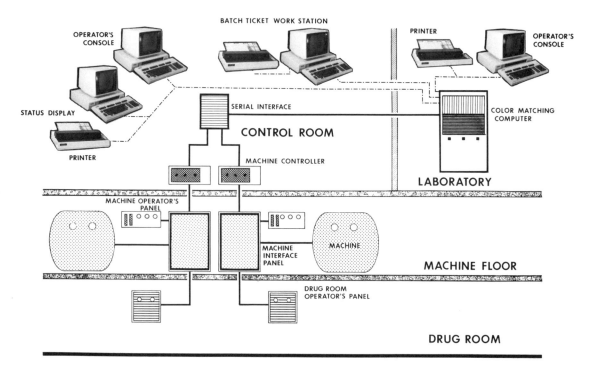

Figure 1.1   Computer controlled textile dyehouse. (Courtesy of Gaston County Dyeing Machine Co.)

stallation. Most of the valves and sensors are digital controlled and others are analog controlled. To illustrate the difference between digital and analog control valves, digital valves have only two states: OPEN or CLOSED. Analog valves can be varied from completely CLOSED to full OPEN or any position in between. To give an example of digital-controlled valves and sensors, suppose that you wanted to fill the keir, and the keir top and drain valve are open. The OPEN sensor inputs from the safety interlock and the drain valve sensors would inhibit the opening of the fill valve and pump until the keir top and drain valve are closed. An OPEN safety interlock would also inhibit the heat exchanger, air pad valve, or vacuum fill valve as a safety measure because some dye procedures require the keir to be pressurized to 100 psi to enable the dye liquor to be heated to a temperature of 149 degrees Celsius (°C) [300 degree Fahrenheit (°F)] without boiling or turning to steam.

Figure 1.5 further illustrates sensor inhibit control by showing a portion of a sensor-inhibit system of a textile dyeing machine. When a valve is CLOSED, its sensor transmits a logic $\emptyset$ to a logic NAND gate; a logic 1 if it is OPEN. A NAND gate must have a logic $\emptyset$ on inputs A and B to output a logic 1 at output C. The four sensor inputs, at the upper left corner of Fig-

Figure 1.2 Batch kiers. (Courtesy of Gaston County Dyeing Machine Co.)

Figure 1.3 Sequence dyeing machines. (Courtesy of Gaston County Dyeing Machine Co.)

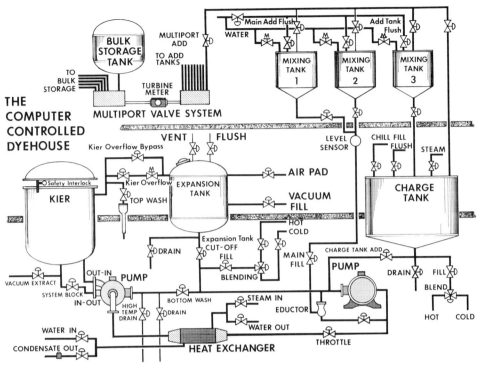

Figure 1.4 Dye machine controlled valves. (Courtesy of Gaston County Dyeing Machine Co.)

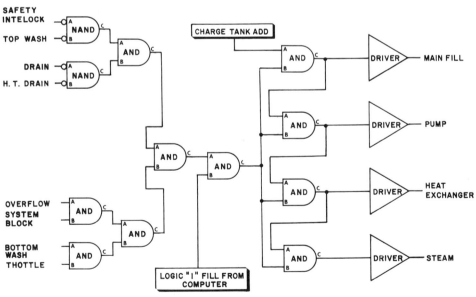

SENSOR INHIBIT CONTROL

Figure 1.5 Inhibit logic circuit schematic drawing.

ure 1.5, must be logic $\emptyset$'s or the system will be inhibited. The four sensor inputs to AND gates, at the lower left corner of Figure 1.5, must have logic 1's at the inputs A and B to output a logic 1 at C. After all critical valves have been set to the required open or closed position, the system will still be inhibited until the computer transmits a logic 1 to the fill-control AND gate. The sensor-inhibit logic circuits are located in the machine control panel and are not an integral part of the control computer.

THE MICROMONITOR COMPUTER

The MicroMonitor II computer, shown in Figure 1.6 in its desktop cabinet, is constructed on a standard 19-in. EIA rack mount panel. Its back panel is equipped with RS-232 serial and parallel connectors for monitor and printer cables and a special control panel to output control signals to the machine interface panel. It has provisions for 96 digital I/O ports, 12 analog input and 4 analog output I/O ports. In addition, there is an on-board random access memory (RAM) of 192k bytes, and controller cards to interface the computer to as many as 8 cathode ray tube (CRT) monitor and printer stations. Each can simultaneously display or print out the dye machine's operation status, logistics data, clock, and calendar, and sound an audible alarm in case of difficulty with the machine. A Winchester disk drive and disk controller card offer the user as much as 5 megabytes of data storage.

Figure 1.6   MicroMonitor computer. (Courtesy of Gaston County Dyeing Machine Co.)

THE MICROMONITOR II FUNCTIONS

The key to computer control lies in the software or program that enables the computer to execute the control commands by directing control interfaces to perform a given action, at a given time, and relays the completion of that control action back to the host computer. The MicroMonitor II Process Control System's software is designed specifically for textile dyeing machine operation and follows the nomenclature of the dye machine shown in Figure 1.1.

Like all software, the MicroMonitor II software begins with a menu of the software contents. Let's examine the MicroMonitor menu and the procedure commands of a typical dye process and the additional control panels that control the dye machine to the programmed functions of the computer.

Main Menu

The menu of the control software displays the following:

1. PROCEDURE EDITOR
2. DISPLAY PROCEDURE FILES
3. DELETE PROCEDURE
4. LOAD PROCEDURE FOR EXECUTION
5. DISPLAY MACHINE STATUS
6. PRINT A PROCEDURE
7. PRINT REPORTS

Procedure Commands

1. Procedure number
2. Type number (procedure identifier for machine type and size)
3. Start cycle
4. Air pad
5. Overfill
6. Fill to level 1 at desired temperature
7. Ramp at desired ending temperature
8. Load
9. Hold for timed period
10. Unload
11. Sample
12. Wash for time period and temperature

13. Drain

14. End of cycle

15. Tangle alert

16. Machine add (manual side of add tank)

17. pH sample

18. Temperature set point

19. Add tank 1 or 2 at a fast or slow rate

20. Prepare tank 1 or 2

21. Fill with programmed quantity and at desired temperature

22. Wash with programmed quantity and at desired temperature

23. Add programmed brime quantity at fast or slow rate

24. Cool wash at desired rate to end temperature

25. Spray for use at boil-out

THE MICROMONITOR CONTROL PANELS

The MicroMonitor II computer is the main control and logistics brain of a number of control units or panels that are located at specific locations at the dye machine, the dye control station, and the drug room. The functions of these panels are as follows:

Main Control Panel

The main control panel (MCP) receives data from all machine sensors, panel switches, and dye operation procedures through its RS-232 or RS-422 serial I/O ports and provides the necessary outputs for machine control. Its standard EIA 19-in. panel is rack mounted in an enclosure designed for use in a control room environment. The integral CRT and keyboard permit stand-alone operation without the need of a supervisory host computer. The MCP panel is shown in Figure 1.7.

Machine Interface Panel

The machine interface panel (MIP) contains the electrical and pneumatic hardware necessary to interface the main control panel to the dyeing machine and the drug room operator's panel. A four-wire RS-232 or RS-422 serial cable and power cable connect the main control panel to the machine interface panel. The MIP contains 96 digital I/O ports and 8 analog input and output I/O ports. Its standard EIA 19-in. panel is housed in a wall-mount enclosure designed for use in a dye machine environment. The MIP is shown in Figure 1.8.

Figure 1.7 MCP panel. (Courtesy of Gaston County Dyeing Machine Co.)

Figure 1.8 MIP panel. (Courtesy of Gaston County Dyeing Machine Co.)

Machine Operator's Panel

The machine operator's panel (MOP) contains the necessary alarms, switches, and readouts for communications with the machine operator. The operator will be alerted for those functions requiring manual intervention, such as load, add, overfill, sample, or unload. Operator acknowledgment of completion of the manual function is required before the computer program will continue. The MOP's EIA 19-in. panel is housed in a pedestal-mounted stainless steel enclosure designed for use in chemical or high-moisture environments. The MOP is shown in Figure 1.8.

Drug Room Operator's Panel

The drug room operator's panel (DROP) serves to alert the operator when the program commands the preparation of a given mix tank. Acknowledgment switches enable the operator to signal the controller when each mix tank is ready. The DROP contains solenoid valves for computer program control of the mix-tank add process. The DROP's EIA 19-in. panel is mounted in a stainless steel enclosure designed for use in a corrosive environment. The DROP panel is shown in Figure 1.9.

Figure 1.9   MOP panel. (Courtesy of Gaston County Dyeing Machine Co.)

INFOR/TEX*

The INFOR/TEX software enables the central computer to prepare the managerial logistics daily reports in the following areas:

Inventory Control

Through INFOR/TEX, the user can monitor and control the current raw material inventory by posting orders and receipts on the CRT terminal of the MicroMonitor II. Since INFOR/TEX maintains accurate information on inventory at all times, management can obtain reports on demand to remain informed on inventory status on hand and committed. Included in the reports are the inventory levels and values of each raw material and an indication of which raw material requires reordering.

Vendor Information

The INFOR/TEXT software allows the MicroMonitor II to keep accurate and timely records of materials ordered from various vendors and the amount purchased from each. When it is time to reorder, the purchasing department can employ INFOR/TEX to determine which vendor to contact and the vendor's address or telephone number.

Formula Maintenance

Most textile dyehouses maintain hundreds to thousands of formulations for coloring, bleaching, and routine dye processing. The INFOR/TEX software enables the MicroMonitor II to maintain all formulations and all procedures required to use the formulations in production. It also monitors formula performance in production and determines formula cost. INFOR/TEX can also be used to determine the quantity of each dyestuff to assure repeatability in color matching to any formula.

Production Control

INFOR/TEX provides a simple way of producing batch tickets complete with formulation and procedures for a dye production run. This simplifies the writing of batch tickets and minimizes errors from calculations or transcribing formulations in manually written batch tickets. INFOR/TEX can be used to monitor production by supplying reports of work in process and production completed by each shift.

*INFOR/TEX is a registered trademark of Applied Color Systems, Inc.

Cost Reduction

From the foregoing INFOR/TEX software report capabilities, cost reduction can be seen in many ways:

1. Error reduction
2. Reduction in clerical labor
3. Increased inventory efficiency
4. Increased production efficiency
5. Increased purchasing efficiency
6. Increased management efficiency

To illustrate how clerical cost can be drastically reduced, Figure 1.10 shows a typical logistics information clerk's computer workstation. One person, with the aid of INFOR/TEX's software, replaces an inventory clerk, a material control clerk, a production control clerk, and a purchasing clerk. One person with the aid of a computer printer can generate more reports on the operation of a textile dyeing facility in 8 hours than can be accomplished in the same period by eight clerks and a typist. Figure 1.11 shows the system's printer as it produces a hard copy of a current production status report.

Figure 1.10 DROP Panel. (Courtesy of Gaston County Dyeing Machine Co.)

Figure 1.11   Logistics clerk at computer station. (Courtesy of Gaston County Dyeing Machine Co.)

To further illustrate INFOR/TEX's capabilities in producing production, inventory, and purchasing assist reports, Figure 1.12 shows samples of raw material inventory, inventory value, and vendor information for reorders. Figure 1.13 shows samples of work in process, work completed, and machine efficiency.

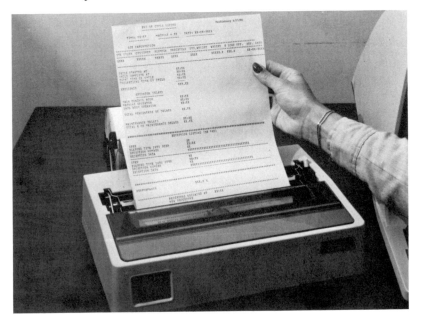

Figure 1.12   MicroMonitor's system printer. (Courtesy of Gaston County Dyeing Machine Co.)

REPORT GENERATION

EXCEPTION REPORT RUN ON 07-DEC-82 AT 11:30   PAGE 1

GREIGE	FORMULA	PROCESS NUMBER	LOTS DYED	STANDARD ADDS/LOT	AVERAGE ADDS/LOT	STANDARD REDYES/LOT	AVERAGE REDYES/LOT	AVG MATERIAL COST STANDARD	AVG ACTUAL	AVG CYCLE TIMES STD	ACTUAL
GR1567	CK1425	P2345	2	1.00	1.53	0.00	0.50	234.56	267.25	3:20	3:55
	CK1547	P235A	9	1.50	1.45	0.00	0.00	145.23	143.23	2:20	2:35

ACE TEXTILE REPORT GENERATION   PAGE 1
** WORK COMPLETED **   LIST OPTION: ALL TICKETS   RUN ON 07-DEC-82 AT 11:00

TICKET ID	FORMULA ID	FORMULA DESCRIPTION	GREIGE ID	GREIGE DESCRIPTION	DATE SCHEDULED	DATE COMPLETED	NO. ADDS	ACTUAL BATCH SIZE	MACH NO.
L2453	CK1783	SUNBIRD YELLOW 1089	GR567	WOVEN POLYESTER	12-NOV-82	19-NOV-82	1	1345.	14
L3421	CK2189	STEEL BLUE 3245	GRB972	POLY DBL KNIT	13-NOV-82	17-NOV-82	2	2387.	8

ARMSTRONG WORLD REPORT GENERATION   PAGE 1
** MACHINE EFFICIENCY DETAILED **   LIST OPTION: SPECIFIC MACHINE BY DATE   RUN ON 07-DEC-82 AT 11:30

MACHINE NUMBER	TICKET ID	FORMULA ID	FORMULA DESCRIPTION	GREIGE ID	GREIGE DESCRIPTION	PROCESS NUMBER	NUMBER OF ADDS	MACH. EFF
14	L1256	CP645A	GOLDEN HAZE 6011	BR1245	NYLON KNIT LOT 6	P1258	1	50%

WEIGHT	YARDS	START - FINISH TIME	START - FINISH DATES	CYCLE TIME	STD TIME
454.	1200.00	1:30 4:45	21-NOV-82 21-NOV-82	3:15	1:50

REPORT GENERATION

EXCEPTION REPORT RUN ON 07-DEC-82 AT 11:30   PAGE 1

GREIGE	FORMULA	PROCESS NUMBER	LOTS DYED	STANDARD ADDS/LOT	AVERAGE ADDS/LOT	STANDARD REDYES/LOT	AVERAGE REDYES/LOT	AVG MATERIAL COST STANDARD	AVG ACTUAL	AVG CYCLE TIMES STD	ACTUAL
GR1567	CK1425	P2345	2	1.00	1.53	0.00	0.50	234.56	267.25	3:20	3:55
	CK1547	P235A	9	1.50	1.45	0.00	0.00	145.23	143.23	2:20	2:35

ACE TEXTILE REPORT GENERATION   PAGE 1
** WORK COMPLETED **   LIST OPTION: ALL TICKETS   RUN ON 07-DEC-82 AT 11:00

TICKET ID	FORMULA ID	FORMULA DESCRIPTION	GREIGE ID	GREIGE DESCRIPTION	DATE SCHEDULED	DATE COMPLETED	NO. ADDS	BATCH SIZE	NO.
L2453	CK1783	SUNBIRD YELLOW 1089	GR567	WOVEN POLYESTER	12-NOV-82	19-NOV-82	1	1345.	14
L3421	CK2189	STEEL BLUE 3245	GRB972	POLY DBL KNIT	13-NOV-82	17-NOV-82	2	2387.	8

ARMSTRONG WORLD REPORT GENERATION   PAGE 1
** MACHINE EFFICIENCY DETAILED **   LIST OPTION: SPECIFIC MACHINE BY DATE   RUN ON 07-DEC-82 AT 11:30

MACHINE NUMBER	TICKET ID	FORMULA ID	FORMULA DESCRIPTION	GREIGE ID	GREIGE DESCRIPTION	PROCESS NUMBER	NUMBER OF ADDS	MACH. EFF
14	L1256	CP645A	GOLDEN HAZE 6011	BR1245	NYLON KNIT LOT 6	P1258	1	50%

WEIGHT	YARDS	START - FINISH TIME	START - FINISH DATES	CYCLE TIME	STD TIME
454.	1200.00	1:30 4:45	21-NOV-82 21-NOV-82	3:15	1:50

MACHINE NUMBER	TICKET ID	FORMULA ID	FORMULA DESCRIPTION	GREIGE ID	GREIGE DESCRIPTION	PROCESS NUMBER	NUMBER OF ADDS	MACH. EFF
NNN	NNNNNNNN	XXXXXXXX	X-----------------X	X------X	X--------------------X	XXXXXXXX	NN	NN%

WEIGHT	YARDS	START - FINISH TIME	START - FINISH DATES	CYCLE TIME	STD TIME
NNNNNNN.NN	NNNNNNN.NN	NN:NN NN:NN	XX-XXX-XX XX-XXX-XX	NN:NN	NN:NN

Figure 1.13   Sample printouts of inventory control forms. (Courtesy of Applied Color Systems.)

15

** RAW MATERIAL INVENTORY LEVEL **

ACE TEXTILE INC.   REPORT GENERATION
LIST OPTION: ALL

PAGE 1
RUN ON 07-DEC-82 AT 15:54

MATERIAL ID	DESCRIPTION	TYPE	UNITS ON-HAND	INVENTORY UNITS	COMMITTED TO PROD.	UNITS ON-ORDER	REORDER POINT	BELOW MINIMUM	CONSUM PER 1	CONSUM PER 2	CONSUM PER 3
DR1310.	RES RED 3GL PST	D	645.	GALLONS	45.	345.	500.		65.	770.	1110.
DR216D	RES RED 2GL PDR	D	25.	POUNDS	0.	75.	100.	*	0.	0.	

** RAW MATERIAL INVENTORY VALUE **

FORRESTAL INC.   REPORT GENERATION
LIST OPTION: RANGE

PAGE 1
RUN ON 07-DEC-82 AT 11:30

MATERIAL ID	MATERIAL DESCRIPTION	UNITS ON-HAND	INVENTORY UNIT	LATEST COST/UNIT	WEIGHTED COST/UNIT	STANDARD COST/UNIT	VALUE AT LAT-COST	VALUE AT WEI-COST	VALUE AT STD-COST
DR131P	RES RED 3GL PST	645.	GALLONS	9.720	8.930	9.000	6269.	5760.	5805.

** VENDOR LISTING **

ARMSTRONG WORLD REPORT GENERATION
LIST OPTION: RANGE

PAGE 1
RUN ON 17-JUN-82 AT 11:30

VENDOR ID	VENDOR NAME AND ADDRESS	SALES CONTACT PHONE NUMBER	RAW MATERIAL ID	RAW MATERIAL DESCRIPTION
CA5362	ABC DYESTUFFS INCORPORATED 23 MAIN STREET P.O. BOX 1234 CHARLOTTE, NORTH CAROLINA 28205	MR. JAMES CHANCE 704-333-1234	RM1234 RM1238 RM2333	DISP RED JG DISP BLUE LS DISP RED MS

** RAW MATERIAL INVENTORY LEVEL **

ACE TEXTILE INC.   REPORT GENERATION
LIST OPTION: ALL

PAGE 1
RUN ON 07-DEC-82 AT 15:54

MATERIAL ID	DESCRIPTION	TYPE	UNITS ON-HAND	INVENTORY UNITS	COMMITTED TO PROD.	UNITS ON-ORDER	REORDER POINT	BELOW MINIMUM	CONSUM PER 1	CONSUM PER 2	CONSUM PER 3
DR1310.	RES RED 3GL PST	D	645.	GALLONS	45.	345.	500.		65.	770.	1110.
DR216D	RES RED 2GL PDR	D	25.	POUNDS	0.	75.	100.	*	0.	0.	

** RAW MATERIAL INVENTORY VALUE **

FORRESTAL INC.   REPORT GENERATION
LIST OPTION: RANGE

PAGE 1
RUN ON 07-DEC-82 AT 11:30

MATERIAL ID	MATERIAL DESCRIPTION	UNITS ON-HAND	INVENTORY UNIT	LATEST COST/UNIT	WEIGHTED COST/UNIT	STANDARD COST/UNIT	VALUE AT LAT-COST	VALUE AT WEI-COST	VALUE AT STD-COST
DR131P	RES RED 3GL PST	645.	GALLONS	9.720	8.930	9.000	6269.	5760.	5805.

** VENDOR LISTING **

ARMSTRONG WORLD REPORT GENERATION
LIST OPTION: RANGE

PAGE 1
RUN ON 17-JUN-82 AT 11:30

VENDOR ID	VENDOR NAME AND ADDRESS	SALES CONTACT PHONE NUMBER	RAW MATERIAL ID	RAW MATERIAL DESCRIPTION
CA5362	ABC DYESTUFFS INCORPORATED 23 MAIN STREET P.O. BOX 1234 CHARLOTTE, NORTH CAROLINA 28205	MR. JAMES CHANCE 704-333-1234	RM1234 RM1238 RM2333	DISP RED JG DISP BLUE LS DISP RED MS

Figure 1.14 Sample printouts of production control forms.
(Courtesy of Applied Color Systems.)

16

THE PROGRAM CONTROL OPERATION

Figure 1.14 contains letter identification to enable the reader to relate the procedure commands of the computer to the control function of the dye machines valves by the MIP. The following example of how the computer procedure commands would control a dye process is not an actual dye process program but one showing sequence action as an illustration.

ID	Procedure command	Function	Control action
a	1	Process number	P1258
b	2	Machine number	14
c	3	Start	Signal MOP
d	4	Air pad	OFF
e	5	Overfill	OPEN
f	6	Fill to level 1 at desired temperature	Signal MIP
g	20	Prepare ADD tank 1	Signal DROP
h	16	Machine ADD (manual)	Signal DROP
i	8	Load	Signal MOP
j	18	Temperature set point	Set 300°F
k	13	Drains	CLOSE
l	4	Air pad	CLOSE
m	21	Fill at 300°F	Pump ON Heat exchanger ON Interlock OFF
n	19	Add tank 1	AT-1 OPEN
o	9	Hold for 120 min	Start timer
p	7	Ramp to 75°F	Heat exchanger OFF COLD water ON
q	4	Air pad	CLOSED
r	11	Sample	Signal MOP
s	12	Wash for 30 min at 120°F	Pump ON Blend ON
t	17	pH sample	Signal MOP
u	13	Drain	Keir drain OPEN Pump OFF
v	10	Unload	Signal MOP
w	14	End of cycle	Signal MOP

MICROMONITOR PROCESS CONTROL ANALYSIS

a, b, c   The process number, machine type, and machine number are information instructions to the machine operator and drug room operator as to what dyes and machine will be used for the current production run. These are not control functions.

d   The air pad valve is a normally closed valve. The keir must be closed and all drains closed before it can be opened. Only temperatures above 212°F require the use of the air pad.

e   The expansion tank has three overflow valves: the keir overflow bypass, keir overflow, and a manual overflow. The manual overflow valve is operator controlled and if closed, the computer can signal the machine interface panel to open the bypass overflow valve.

f   This command sets up the machine interface panel to control the liquid level and temperature that the process will require on the FILL command (m).

g   This is a manual function of the drug room operator. On a signal from the computer, the drug room operator adds the dye compounds to the mix tank in the amounts specified by the computer to obtain the correct color match that the production run requires.

h   When the drug room operator has completed adding the dye compounds to the mixing tank, the computer sends a signal to alert the operator to open the manual ADD valve so that the computer can control the actual add function when called for by the program.

i   Load is another machine operator manual operation. On a signal from the computer to load, the computer program will be inhibited until the operator signals that the load function has been completed.

j   The computer receives the temperature set point from the process program and sends the machine interface panel the data for reference to the analog temperature sensor feedback to the MIP. Once the temperature of the keir matches the set point temperature, further heating of the keir dye liquor is inhibited.

k   The computer will be inhibited from performing the fill function (m) until all system drains are closed. When the drain close signal is received, the MIP automatically checks all sensor inputs for logic $\emptyset$ and closes any drain valve that might be open.

l   The program calls for a dye liquor temperature of 300°F and is further inhibited until the keir vent is closed. When the temperature of the fill reaches the point where pressure is required in the keir, the air pad valve is opened to raise the pressure to the programmed level.

m   When all systems checks of valves are in their proper state, the fill function is actuated and the dye program resumes.

n   The computer automatically opens the ADD valve as required by the program. However, should the add tank's manual valve be closed or the add compounds not completed, the computer will be inhibited until the drug room operator signals that both requirements have been completed.

o   The hold time can consist either of circulating the dye liquor in and out of the keir for the time period or it can be a filled keir and the fabric or yard being dyed held in a soak period. In either case the program cannot move ahead until the preset hold time has lapsed.

p   This instruction calls for the dye liquor to be cooled down gradually over a period of time, from the maximum temperature to the preset ending temperature called by the process schedule.

q   The air pad valve will be computer closed by the MIP when the temperature of the dye liquor falls below the preset point. The keir's vent valve must be opened to reduce the keir's pressure to atmosphere level before the sample instruction can be sent to the machine operator.

s   Some processes call for a wash period after the sample manual function. The keir does not have to be under pressure again unless the wash temperature must go higher than 200°F.

t   The pH command is another manual function that calls the computer to signal the machine operator to measure the pH of the remaining wash water.

u   On completion of the dye schedule, the low-temperature drain and the expansion tank drain are opened to remove all remaining water from the system.

v   When the liquid-level sensors indicate that the system is empty of liquids, the computer sends a signal to the machine operator to unload the keir. When the keir has been unloaded, the operator signals the computer and an end-of-cycle return signal completes the production run. The machine is now available for another production cycle.

COMPUTER-CONTROLLED MACHINE SHOP

At times it is difficult to draw a line between a computer-controlled machine and a computer-controlled process. Take as an example a machine shop where a computer controls the metalworking machines and the steps of cutting, drilling, milling, and turning operations required to produce a finished metal product. Here the computer controls both the machine and the process, as it does with the textile dyeing machine and the dye process.

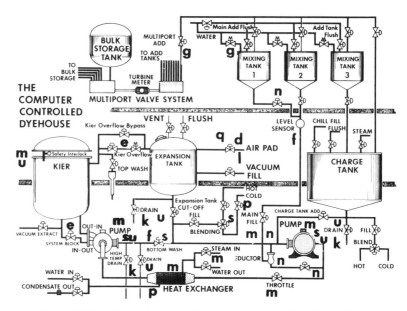

Figure 1.15   ID overlay keyed to program function descriptions.

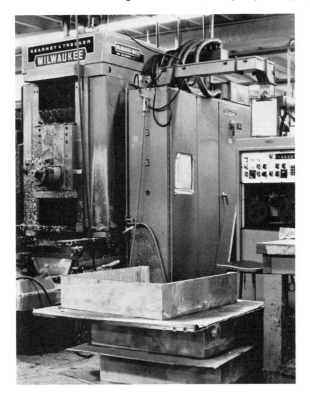

Figure 1.16   Kearney Trecker Model 180 milling machine.

Some industrial cutting machines perform only a single operation, with the machine and cutting operation controlled by magnetic tape or a combination of computer program and tape. In this instance, there is no requirement that the machine be able to interchange cutting tools for different operations (see Figure 1.15). In cases where the controlled machine must perform a variety of cutting operations and control a number of different cutting tools, the required tool may be operator exchanged on a signal from the computer or automatically exchanged by means of a rotating toolholder turret (see Figure 1.16). Figure 1.17 shows a fully controlled metal lathe where the computer program not only selects the correct cutting tool by rotating the toolholding turret, but automatically advances the bar stock in the collet chuck. This machine's operator loads the bar stock on a conveyer bed behind the collet chuck and removes the finished part after the computer completes the operation and severs the part from the bar stock. Figure 1.18 exhibits still another type of computer-controlled metal lathe in which metal castings are loaded onto an adjustable chuck and the computer controls the machine to perform a number of different drilling operations by automatically rotating the drill-bit holding tool turret for each hold diameter required.

Figure 1.19 shows the machine operator's computer and machine control cabinet, which is very similar to the machine operator's panel at the textile dyeing machine. The major difference is that the textile computer signals are basically digital except for the analog control signals, temperature and

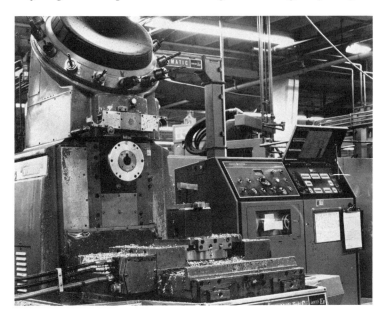

Figure 1.17   Milwaukee Matic EB.

Figure 1.18   Jones and Lampson Economizer automatic lathe.

Figure 1.19   LeBlond Regal 24 automatic lathe with GE controller.

Figure 1.20   Kerney and Trecker machine operator's panel and machine interface control cabinet.

pressure. The cutting machine control signals are basically analog in order to control the speed of rotation of the chuck or cutting tool as well as the X-Y axis for positioning the cutting tool and limiting the depth of the cut or drilling. Generally, the only digital control signal commands are for stop/start functions. Figure 1.20 also shows the tape data storage system required in locations too hostile for disk-drive operation. Machine shops not only have moisture and oil contaminants around the computer control station, but the many cutting operations contribute fine metallic dust particles that would destroy a disk drive's read/write head and the magnetic disk.

CHAPTER REVIEW QUESTIONS

1. What dyeing machine function was the first to be automated?
2. What discovery was the first major improvement in the art of dyeing?
3. What is an electropneumatic device?
4. What is the difference between AND and NAND gates?
5. Define the difference between analog and digital signals.
6. How many control panels are required to operate a textile dyeing machine with the MicroMonitor II computer?
7. To which control panel are the sensor status signals from the dye machine valves directed?

8. Name two dyeing machine functions that must be analog controlled.

9. Does the computer-controlled dye machine and dye process eliminate all manual operations in the control program?

10. What manufacturing functions are performed by the INFOR/TEX software?

11. Identify three machine valves that could inhibit a FILL command if the valves were incorrectly set.

12. On a FILL command, should the air pad valve be open or closed?

13. Why must a keir be pressurized for a dye process?

14. Give three operator functions that are manually performed in a fully automated textile dye process.

15. What is the difference in computer control signals between textile machines and metal-cutting machines?

16. Why are disk-drive data storage units usually not incorporated in the machine operator's computer console or panel?

GLOSSARY

Analog. An electrical signal that varies smoothly, or is continuously variable; analog functions include temperature, pressure, light, and sound.

AND gate. A discrete logic circuit requiring a logic 1 on each input point to produce a logic 1 on the output; an AND gate is comparable to two single-pole switches wired in series, which would require each switch to be closed before current could flow.

Automation. A system or method of operating a machine or process automatically by means of electric or electronic devices without human intervention.

Bar stock. Metal in the form of square or round bars from which metal parts are produced by metal-cutting machines.

Binary. A number system to the base 2 used in connection with computer arithmetic and logic circuit functions.

Bleach. To lighten or reduce the natural tincture or color of yarns and fabrics prior to dyeing to a specific color.

Byte. An 8-bit binary number representing an alphanumeric computer word; denoting the capacity of RAM or disk data storage.

Cam. An irregular shape or contour disk or cylinder which when rotated imparts a rocking motion to any device in contact with the cam's contour.

Collet. A compression-type lathe chuck for quick release of round bar stock.

Digital. An electrical signal that changes abruptly or in steps; digital forms include money, pulse rate, pages in a book, or a typewriter; digital action can be STOP or GO, ON or OFF, YES or NO.

EIA. Electronic Industry Association.

Electropneumatic. A combination device by which electrical or electronic analog signals control a mechanical component used to control the flow of air or a device operated by compressed air.

Inhibit. To check, arrest, stop, or prevent an action from taking place.

Integral. An essential part of a complete assembly or unit.

Interface. Computer or electronic devices designed to work or communicate from one system to another by circuits that isolate or make compatible the communications differences.

Keir. A large vat in which fibers, fabrics, or yarns are boiled, bleached, or dyed.

Kilo. A metric term denoting the number 1000 generally identified by the lowercase letter k; in computer applications, k denotes the number 1024; the MicroMonitor's 192k RAM is actually 196,608 bytes of memory.

Liquor. A liquid medium containing bleach or tinctorial substances used in a textile dye process.

Logic states. Digital logic states are commonly referred to as logic 0 or logic 1. In positive digital logic, the logic 0 is represented by low or no positive voltage, whereas the logic 1 is represented by a positive voltage ranging from 3.25 to 5 V dc.

NAND gate. The not AND or NAND gate logic circuit operates similarly to the AND gate by requiring both inputs to have the same input logic state to output a logic 1. However, the NAND gate requires a logic 0 on each input instead of the logic 1 required by an AND gate.

Software. Computer program data, stored on magnetic disk or tapes, which enable the computer to perform procedural functions such as inventory, procurement, and word processing without creating a new program each time a function is required.

Tinctorial. The part of a solution that produces color, as in bleaching and dyeing of fabrics and yarn.

Turret. A rotating device containing holders for tools as used in the metalworking industry.

Vat. A large container, such as a tub or tank, for holding liquids.

2

Computer Control
in a Hostile Environment

If anything can go wrong, it will regardless of what you may do to prevent it.
Murphy

The computer is one of the most sophisticated electronic devices ever developed. Its complex and delicate electronic circuits make it extremely sensitive to its environment and highly susceptible to data pollution from a multitude of contaminants and electrical interference.

CONTAMINANTS HAZARDOUS TO COMPUTERS

Contamination exists in many forms: atmospheric contaminants, operator contaminants, nature-produced interference, and electrical device interference.

Class I

Atmosphere Contamination

1. Airborne particles, such as dust, lint, fibers, and pollen
2. Atmosphere components, such as humidity, temperature, and electrical discharge (lightning and static electricity)

Human Contamination

1. Contact contamination from hair, hands, makeup, and clothing
2. Contamination from eating, chewing, drinking, and smoking

Class II

Electromagnetic Contamination

1. Radio-frequency interference (RFI) from AM, FM, TV transmitters, commercial two-way radio, radio telephones, citizen's band or amateur transmitters, and hobby radio control transmitters

Electromagnetic Interference

2. Radiated appliance interference from electric shavers, microwave ovens, power tools, vending machines, television receivers, other computers, calculators, automobile ignition, and any other gasoline motor–powered tool or machine

Class III

Electrical Interference

1. Electrical-power-line interference from air conditioners, furnaces, refrigerators, electric motors, fans, flashing electric signs, electric stoves, arc welders, and power disconnects

Class IV

Physical Hazards to Cables

1. Chemical hazards, such as contact with acids, alkalies, and corrosive gases
2. Abrasive hazards, such as friction damage from moving contact with sand, gravel, stone, concrete, brick, glass, wood, or metal
3. Crushing hazards, such as damage from vehicle wheels, stacking of heavy objects, machine contact with cables, and closing of fire or passageway doors

Class I contaminants seldom produce adverse operating conditions to the computer itself. However, they are the most frequent cause of problems with computer keyboards and magnetic data storage peripherals in the form of cassette recorder/players, reel-to-reel recorder/players, and disk-drive units, both floppy and hard disk types. The disk-drive units are most susceptible to particle contamination because of the very small clearance between the disk read/write head and the data disk. Even the 0.25-micron size (μm) of the solid particles in the smoke of a cigarette can produce data pollution in the high-density disk drives.

A temperature- and humidity-controlled environment is important to the operation of the control computer and its magnetic data storage peripherals. It provides the best atmosphere to assure proper cooling of the computer circuits and storage of magnetic tape or disk. If the computer is not properly cooled, there is danger of thermal runaway developing, with destruction of the computer's logic and memory integrated circuits.

The lack of humidity control produces a condition whereby static electricity can be generated by walking across a carpeted floor. Static electricity is extremely hazardous to magnetic data tapes or disk and the complementary metal-oxide semiconductor (CMOS) structure of the integrated circuits inside the computer.

Before leaving this discussion of the effects of contaminants, we should mention a potentially disastrous threat to the computer and its peripherals from spike voltages occurring on utility power lines. Electrical contaminants, class III, include a number of electric motor–operated appliances and devices which are characteristically producers of short-duration power-line spike voltages. Although motors are the prime cause of power-line spike voltage, any high-current device can produce voltage spikes.

The cause of power-line voltage spikes is a characteristic of current moving through conductor (wire), which produces a magnetic field around the conductor. The moment that current begins to flow through the conductor, a magnetic field begins to expand outwardly in all directions parallel to the wire. If the current flow is halted abruptly, such as by cutting off the power with a switch, the magnetic field surrounding the conductor will collapse instantly back into the conductor and inductively generate a high-voltage spike. The spike voltage is dependent on the amount of current flow and the length of the conductor. The collapse of the magnetic field can produce a spike voltage ranging from 600 to 2000 V for a period of several milliseconds. Motor-driven appliances tend to generate higher spike voltages because the field and armature windings increase the effect of the collapsing magnetic field over the induced voltage of a straight conductor typical of power lines. Electric motor appliances also contribute radiated interference because of the arcing of the motor's armature brushes while the motor is running. This effect is similar to making every electric motor a miniature spark-gap radio transmitter. Not only does the motor radiate this form of electrical hash, but it also imparts the hash back into the power lines to compound its ability to produce data pollution. Should this happen to a computer without the protection of a power-line spike and interference filter, the spike voltage could be catastrophic to every electronic component in the computer, video terminal, and disk-drive unit.

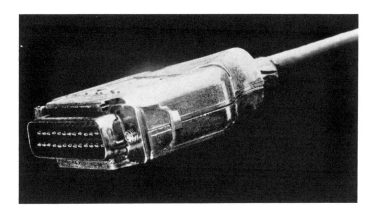

Figure 2.1   RS-232C shielded cable and connector for computer serial connections.

PERIPHERAL DATA LINKS IN A HOSTILE ENVIRONMENT

Business computer peripherals are seldom installed more than 10 ft from the computer. In control applications, the computer peripherals may be located several hundred feet to several thousand feet away and often in a hostile environment. It is obvious that the normal peripheral data link cables (see Figures 2.1 and 2.2) would be totally inadequate as data links for long distances such as those required for industrial control installations. The first recourse would be to fabricate the long data links with cables containing a shield of foil or wire braid surrounding the data-carrying conductors. Unfortunately, improperly installed shielded cables can actually create more EMI interference than the cable was intended to eliminate.

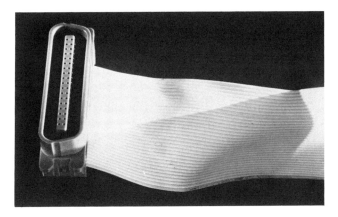

Figure 2.2   Flat ribbon cable and connector for parallel computer connections.

SHIELDED DATA LINKS AND METHODS OF SELECTION*

"SHIELD THAT CABLE"

Bruce Morgen, Associate Editor
Electronic Products

With the FCC clamping down on EMI/RFI emissions from small computer systems, there has been renewed interest in the shielding of cables as a means of holding down system noise generation. Then there is the potentially disastrous problem of data pollution caused by ambient EMI/RFI pickup by inadequate shielding cabling. As frustrating as it is to have an otherwise good system fail to meet FCC requirements, it is equally aggravating to watch your latest design go down every time a car with noisy ignition system goes by! While well-shielded cables are not a cure-all for such maladies, they often make the critical difference between a reliable . . . and salable . . . product and an expensive white elephant.

Shield Performance

If shield effectiveness was the only criterion for selecting shielding, you could simply run wiring through pipes with solid copper walls. Since most of us require a good deal more flexibility of our cables than that of a solid-wall copper tube, we generally settle for less-than-ideal shielding. Instead, we strive for shielding effectiveness that approaches ideal performance as closely as possible within the real-world constraints of flexibility, termination considerations, and cost. To do this, we need a practical way to measure a shield's performance at the frequency involved. While there have been many attempts to arrive at a useful measurement of shielding effectiveness, the current consensus among wire and cable experts favors transfer impedance.

Transfer impedance is the ratio of the potential difference applied at a pair of terminals in a network to the resulting current at a second pair of terminals. If the network in question is a single-conductor shielded cable, terminated with its characteristic impedance and lo-

*Pages 30 through 40 are reprinted in their entirety (including illustrations) by permission of *Electronic Products*, Hearst Business Communications, Garden City, N.Y., Steve Weitzer, Editor, and Bruce Morgen, Associate Editor. Fiber optic data by permission of Belden Fiber Optic Products Division, Geneva, Illinois, Thomas Sykes, Public Relations Manager.

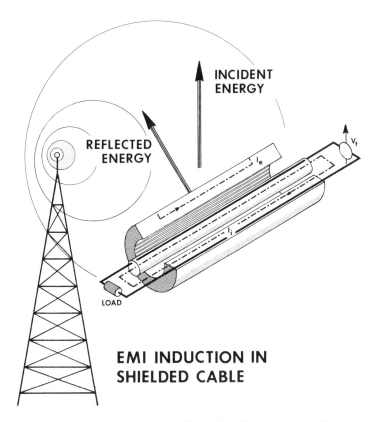

INCIDENT
ENERGY

REFLECTED
ENERGY

EMI INDUCTION IN
SHIELDED CABLE

Figure 2.3 EMI induction in shielded cable. (Reproduced with permission of *Electronic Products*, Hearst Business Publications, and Belden Fiber Optic Products Division.)

cated in an ambient electromagnetic field, an interference current (I_e) is induced in the shield (see Figure 2.3).

What Happens in the Shield?

Part of the incident electromagnetic energy is reflected from the shield and part of it penetrates the shield. This penetrating energy is subject to a degree of attenuation that is dependent on the effectiveness of the shield. Whatever energy does get through generates an interference voltage (Vt) on the cable's circuit and the current flowing through it (Ii). The better the shield, the lower the interference voltage for a given interference current. By calculating the ratio of interference voltage to interference current (Vt/Ie) and controlling the frequency in the electromagnetic field, we can determine the transfer impedance (Zt) of the shield at that frequency.

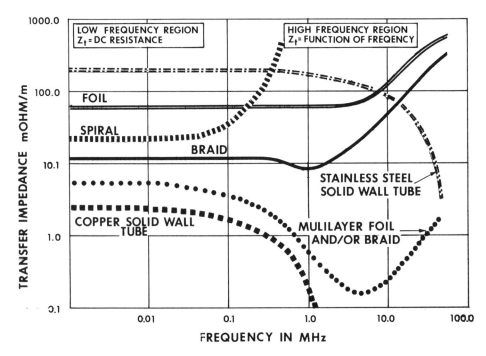

Figure 2.4 Graph of transfer impedance for various shielding methods or types. (Reproduced with permission of *Electronic Products*, Hearst Business Publications, and Belden Fiber Optic Products Division.)

As is apparent from the complete equation . . . $Zt = Vt/Ie$ . . . the lower the transfer impedance, the better the shield. The transfer impedance is expressed in ohms per meter. The transfer impedance in ohms/m is obtained by dividing Zt by the length of the sample, in meters, or by using one meter as a standard length.

Figure 2.4 shows the transfer impedance of some typical shields used for electronic cables, along with our hypothetical solid-wall copper tube and a similar tube made of stainless steel. Below 100 kHz (0.1 MHz) the transfer impedance is equal to the DC resistance. This is why the more conductive copper tube is a better shield than a dimensionally identical stainless steel tube. The other factors influencing shield transfer impedance are thickness and magnetic permeability. As frequency increases, current tends to flow along the inside surface of the shield. This is referred to as "skin effect." Skin effect results in increased attenuation by the shield, which is observable in solid-wall shields as decreasing transfer impedance with increasing frequency.

Another common method of shield evaluation is optical coverage. This is the method used by cable manufacturers when they describe a shielded cable as having "96% coverage." Such figures are arrived at by dividing the surface area of the cable's inner conductor insulation that is covered by the shield by the total surface area of that insulation.

Shield Construction

The most flexible of the commonly available cable shields is the spiral wrapped or served shield. Spiral shields consist of bare or tinned copper strands wrapped around insulated conductor (s). Optical coverage as high as 97% is possible with this construction. The major limitation of spiral shielding is generally poor performance at frequencies above 100 kHz. This is due to the highly inductive nature of the spiral configuration, which produces longitudinal solenoid magnetic fields. These fields add to transfer impedance resulting in a curve that rises steeply with increasing frequency. Good inductors that they are, spiral shields can act as low-pass filters, generally not a desirable trait. On the plus side, spiral shielded cables are easy to terminate and quite low in cost.

Braid Is Better

Somewhat better performance is obtainable with braided shields. The shield is woven with copper (or aluminum) stranded carriers, with one set woven in a right-hand lay and the other in a left-hand lay (see Figure 2.5). Braided shield cables are not as flexible, easy to terminate, or as inexpensive as spiral-wraps. Transfer impedance at low frequencies is

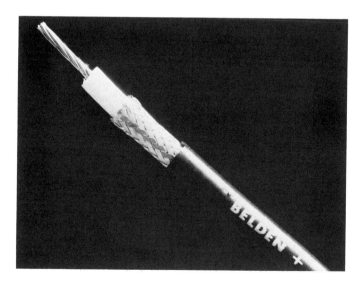

Figure 2.5   Copper braid coaxial cable. (Reproduced with permission of *Electronic Products*, Hearst Business Publications, and Belden Fiber Optic Products Division.)

lower than spiral-wraps because of considerable lower DC resistance . . . the crossing strand carriers shorten the signal path. The transfer impedance does not rise significantly until well above 1 MHz because the oppositely wound strand carriers produce compensating solenoidal fields rather than a single longitudinal field generated by spiral-wrapped shields. The major limitation of braided shielding is that the weave inherently contains apertures. By carefully selecting weave parameters like wire gauge, ends per strand of carrier, picks per unit of cable length, and weave angle, optical coverage as high as 97% are possible. However, this percentage is not sufficient to prevent transfer impedance from increasing to a measurement that can be 20 times higher at 30 MHz than it is at 10 MHz.

The most recent development in cable shielding is the foil shield (see Figure 2.6), which usually consists of aluminum foil laminated to polyester tape or adhesively bonded to the inner conductor insulation. Since the foil is only 1 to 3 microns (0.000039″ to 0.00012″) thick, the shield's DC resistance—and its transfer impedance at low frequencies—is relatively high despite the "drain wire." The foil shield is wound spirally, but the tape is wide enough to minimize inductive effects. The most promising of its virtues is that 100% optical coverage is possible. One would think that the transfer impedance of the foil shield would show a decrease with increasing frequency like the solid-wall copper

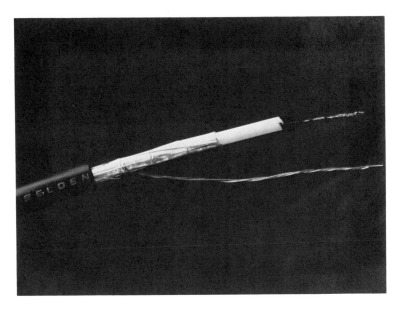

Figure 2.6 Foil shield with drain wire. (Reproduced with permission from *Electronic Products*, Hearst Business Publications, and Belden Fiber Optic Products Division.)

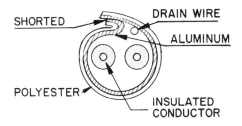

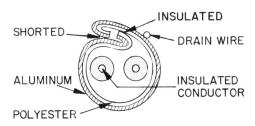

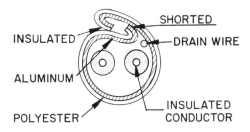

Figure 2.7   Drawing illustrating types of foil shorting folds. (Reproduced with permission from *Electronic Products*, Hearst Business Publications, and Belden Fiber Optic Products Division.)

tube. Instead, first generation foil shields showed a transfer impedance increase similar to that of braided shields. This is due to the lack of electrical continuity across the seams between the turns of foil shield tape. The polyester or adhesive backing, necessary for strength, was acting as an insulator and creating a spiral-shaped aperture or "slot" that allowed electromagnetic field penetration.

Since foil cables are relatively inexpensive and provide easy shield termination via the uninsulated drain wire, there was plenty of motivation for solving the slot problem. The solution arrived at was the shorting fold (see Figure 2.7), which effectively eliminates the slot, at least at frequencies below 10 MHz. Shorting-fold variations have been developed which allow the extra benefit of using polyester backing as an additional layer of insulation, a valuable feature if the foil-shield conductor is part of a crosstalk prone cable assembly.

Foil-shield performance can be further improved by additional drain wires and a second layer of foil on the opposite side of the polyester. Foil-shield cables with multiple drain wires and two layers of foil can have lower transfer impedance than typical braided-shield cables at 15 MHz or higher.

Belt and Suspenders

A significant trend in cable shielding combines the foil and braided approaches (see Figure 2.8). Covering a two-layer foil shield with a braided shield can result in a transfer impedance curve that falls starting at between 1 and 3 MHz . . . performance that begins to resemble that of the solid-wall copper tube. This shielding approach not only out performs the best braided shields, it is also far less costly. Since the foil provides 100% optical coverage, there is no need to minimize braid apertures, and relatively inexpensive braids with optical coverage in the 40 to 80% range can be used.

If even closer approximation of solid-wall copper tube performance is desired, more layers of foil and braid can be added. A good example of this type of approach can be found in the cable used for the Ethernet local area network (see Figure 2.9). Ethernet cable has a polyester-backed foil shield bonded to the inner conductor insulation, covered by a tinned copper braided shield with 92% optical coverage, a two-foil shield sharing a common polyester backing, and a second tinned-copper braided shield with 92% optical coverage. Transfer impedance of this shield configuration falls steadily with increasing frequency up to 30 MHz. Three-layer shields developed for CATV coaxial cables are better yet. These designs use a bonded-foil/braid/foil-with-

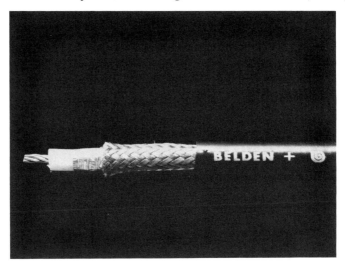

Figure 2.8 Combination of foil and copper braid coaxial cable. (Reproduced with permission from *Electronic Products*, Hearst Business Publications, and Belden Fiber Optic Products Division.)

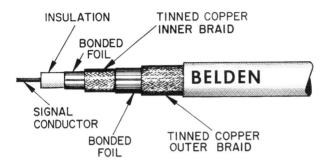

INSULATION   TINNED COPPER
               INNER BRAID
   BONDED
   FOIL

SIGNAL
CONDUCTOR
        BONDED   TINNED COPPER
        FOIL   OUTER BRAID

Figure 2.9 Belden Ethernet four shield coaxial cable. (Reproduced with permission from *Electronic Products*, Hearst Business Publications, and Belden Fiber Optic Products Division.)

shorting-fold shielding format and exhibit transfer impedance curves that continue to fall until between 80 and 100 MHz.

About Emissions

If a designer's major concern is EMI/RFI emissions, rather than susceptibility, the question often arises: Does transfer impedance tell me anything about a shield's effectiveness against emissions from the cable? Since transfer impedance measurements result from the effects of an electromagnetic field originating outside the cable, they essentially measure susceptibility. Moreover, current theoretical formulae for calculating shield effectiveness against cable-sourced EMI/RFI are rather complex and generally not published by the suppliers.

Cable makers do, however, perform carefully controlled radiation tests. When viewed together with the transfer impedance curves of the same shields (see Figure 2.10), these tests show that transfer impedance

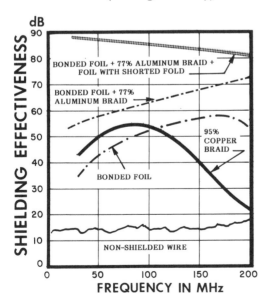

Figure 2.10 Relationship between transfer impedance and shielding effectiveness.

measurements correlate closely enough to shield effectiveness against cable-sourced emissions to be used as a selection guideline.

Shield Connection Formats

Although "grounded" cable shields, with the shield connected to circuit and chassis ground at both ends of the cable (see Figure 2.11) provide sufficient EMI/RFI protection in most applications, there are cases where this approach can lead to ground loop or common mode interference. This is caused by differences in potential between nominal ground points to which opposite ends of the shield are connected. This condition permits unwanted noise to be carried on the shield along with the signal return. This problem, which is especially troublesome at frequencies of 6 MHz or lower, can be remedied by separating the chassis and circuit/shield grounds (see Figure 2.12) and reducing the number of ground connections to the shield to a minimum. This requires that chassis-mounted cable connectors be insulated from panels and connected to the circuit ground.

It has been pointed out that neither aluminum nor copper shielding is effective against low-frequency magnetic fields. In situations where such fields are present, or where very high levels of EMI/RFI or crosstalk are a problem, cables employing electrically separated multiple shields, twisted-pair inner conductors, or a combination of the two configurations are often used.

Although no more effective than the usual coaxial shielded cable configuration against low-frequency magnetic fields, triax cable, with its two shields, provides a number of advantages. By grounding the

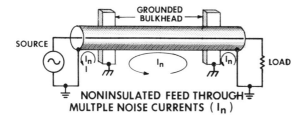

NONINSULATED FEED THROUGH
MULTPLE NOISE CURRENTS (I_n)

Figure 2.11   Shielding methods.

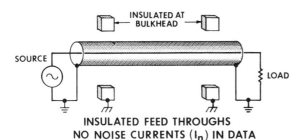

INSULATED FEED THROUGHS
NO NOISE CURRENTS (I_n) IN DATA

Figure 2.12   Shielding methods.

outer shield and using the inner shield exclusively as the signal return (see Figure 2.13), not only are ground loops avoided, but capacitive coupling of external noise fields to the signal carrying conductors is also prevented. Triax can also be connected in a "driven shield" configuration (see Figure 2.14). When used this way, the cables outer, grounded at the source end, is the signal return. The inner shield is connected to the center conductor at the source and acts as a Faraday shield. This has no particular advantage with respect to interference control, but it does reduce the distributed capacitance of the cable considerably. This permits longer cable runs in systems where high data rates are transmitted.

One approach that is effective against low-frequency magnetic fields is to use balanced line transmission via twisted-pair center conductors along with a shield. Twinax cable (see Figure 2.15), with its specified characteristic impedance, is an increasingly popular example of this technique. It is the least expensive cable that is effective against EMI/RFI, crosstalk, and low-frequency magnetic fields. Its major disadvantage is a frequency ceiling of 15 MHz due to signal loss.

By combining triax's two separate shields with twinax's twisted-pair center conductors, quadrax cable acquires the virtues of both. Perhaps the ultimate in EMI/RFI and crosstalk shielding effectiveness results from connecting the outer shield of the quadrax cable to the system ground and the inner shield to a true "earth" ground (see Figure 2.16). Where an earth ground is impractical, similar performance can be had by connecting the inner shield to the system ground at the

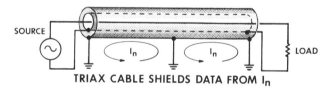

TRIAX CABLE SHIELDS DATA FROM I_n

Figure 2.13   Shielding methods.

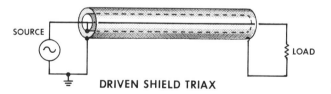

DRIVEN SHIELD TRIAX

Figure 2.14   Shielding methods.

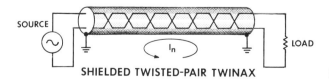

SHIELDED TWISTED-PAIR TWINAX

Figure 2.15   Shielding methods.

SOURCE

LOAD

QUADRAX

Figure 2.16 Shielding methods.

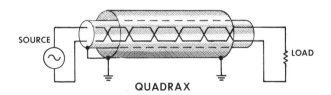

SOURCE

LOAD

QUADRAX

Figure 2.17 Shielding methods.

source end only, while retaining the outer braid ground connections at both ends of the cable (see Figure 2.17). Where the low-frequency magnetic field suppression of the twisted-pair center conductors is not needed, both of these shield configurations can be used with a coaxial cable having three separate shields.

DATA TRANSMISSION VIA BEAMS OF LIGHT

Shielded coaxial cables have long served the communications field as radio-frequency links between the transmitter and the antenna, long-distance transmission of network television programs, and audio coupling links in high-fidelity audio systems. Although satisfactory service is obtainable from multishielded cables, a number of disadvantages remain; they are bulky, heavy, lack flexibility, and have an inherent power-loss factor.

Early in the 1970s, the need for secure communications prompted the military to seek industry's help to develop a method for optical-signal transmission using very small optical fibers. The fiber optic cable or optical waveguide as it is sometimes called, has proved to be the ultimate means for transmission of high-speed data and low-speed process control signals. Here are some of the advantages of fiber optic cables:

1. Electromagnetic interference and crosstalk immunity
2. No electrical ground loops or short circuits
3. Small size and light weight
4. Large bandwidth for size and weight
5. Safe in combustible areas (no arcing)
6. Immunity to lightning and electrical discharges
7. Longer cable runs between repeaters
8. Flexibility and high strength

9. Potential high-temperature operation
10. Nuclear-radiation resistant
11. Secure against signal leakage and interference
12. No electrical hazard when cut or damaged

The advantages of fiber optic cables clearly marks its superiority over all forms of shielded cable for hostile-environmental applications.

INTRODUCTION TO FIBER OPTICS TECHNOLOGY

The basic element of this technology is the optical fiber, a small, transparent fiber that guides optical energy in the form of visible light or infrared radiation. This fiber consists of an inner transparent silica (glass) core which is surrounded by an outer transparent material, commonly referred to as the cladding (see Figure 2.18). The optical fiber and cladding configuration is somewhat similar to a solid-strand wire surrounded with an insulation material or the coaxial cable's shield surrounding the inner insulated conductor. However, the similarity stops when one tries to compare methods of signal transmission by wire and optical fiber. The wire uses analog or digital variations of an electrical current, while the optical fiber uses analog or digital variations of a frequency of light.

Guiding the Optical Signal

Light is guided in the core by reflections at the core–cladding interface. These reflections occur because of a difference in the refractive index (h) between the higher-index core and lower-index cladding. Reflections between high- and low-index media are described as total internal because 100% of the energy is reflected if the interface is smooth and both materials are transparent. The refractive index of a material is a number which relates the velocity of light in a vacuum to that of another material, such as the fiber's core glass. For a fiber's core the index is about 1.46, which means that the optical

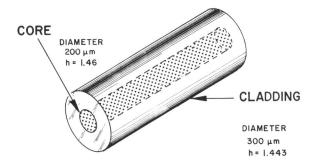

CORE
DIAMETER
200 μm
h = 1.46

CLADDING
DIAMETER
300 μm
h = 1.443

Figure 2.18   The core/cladding configuration of a step index optical fiber. (Courtesy of Belden Fiber Optics.)

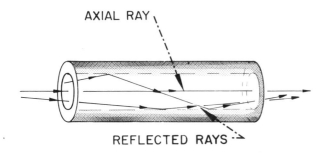

Figure 2.19   Typical light ray propagation through an optic fiber. (Courtesy of Belden Fiber Optics.)

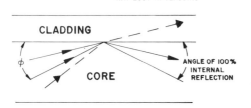

Figure 2.20   Ray paths reflected from the core/cladding interface. (Courtesy of Belden Fiber Optics.)

energy propagates at about 67% of its velocity in air. Calculating the velocity of light in a fiber yields a value of 2.1 times 10 to the eighth power in meters per second, or a propagation time of 5 nanoseconds (ns) per meter. This means that an axial ray will require about 5 microseconds (μs) to propagate a distance of 1 km. The path of typical light rays is depicted in Figure 2.19

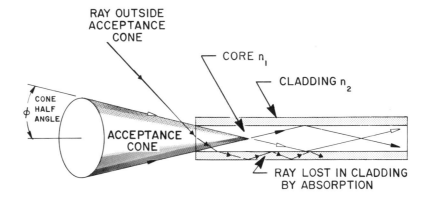

NUMERICAL APERTURE NA

$$NA = SIN\,\phi = \sqrt{n_1^2 - n_2^2}$$

Figure 2.21   Numerical aperture and acceptance cone half-angle. (Courtesy of Belden Fiber Optics.)

as the rays propagate down a step-index fiber. Figure 2.20 further depicts the path of those light rays that reflect at angles less than the acceptance half-angle and those lost by passing through the cladding because their reflection angle is greater than the acceptance half-angle. Acceptance cone definitions and typical values for the numerical aperture are presented in Figure 2.21.

Types of Optical Fibers

There are two basic types of optical fibers: step index and graded index. The term *step index* refers to the index of the core being a step lower or higher than the cladding. There is no gradual change in index of the core to the interface with the cladding as there is in the graded index. There are marked size differences in the core diameters of step and graded indexes. The step-index core ranges from 200 microns (μm) to 300 microns in diameter, while the graded-index core ranges from 50 to 100 μm. There are bandwidth limitations to step and graded fiber cores, with step cores limited to between 20 and 25 MHz, while graded cores range from 100 to over 1000 MHz.

One method by which the step-index fiber is made involves encasing the silica core in a tube of cladding glass, then fusing the two glasses together to make the draw preform. The preform is then transferred to the draw tower, where the fiber is drawn down from the preform by the application of a high temperature of approximately 2000°C and the draw tension. The diameter of the fiber is controlled by varying the draw tension. Increasing the tension reduces the fiber diameter, and decreasing the tension enlarges the diameter.

The pristine glass of the newly drawn fiber must be immediately protected from contaminants, which could cause degradation, and flaws, that could affect the optical and physical integrity of the fiber. Accordingly, a coating of silicone RTV (room-temperature vulcanizing) is applied by passing the fiber through a reservoir containing RTV. After curing the RTV, the fiber will receive an application of a hard buffer plastic by means of a mini-extruder. The final on-line process is the proof 100% elongation test, where the fiber is subject to testing at 100,000 psi. Figure 2.22 presents completed fiber optic cables and shows the high-strength members surrounding the optical fiber.

The limitation of step-index optical fibers lies in the angle paths taken by the light rays as compared to the axial light rays. Naturally, those rays of light that enter at the zero angle of acceptance encounter little or no reflections in their passage through the optical fiber. Those light rays entering the cone of acceptance at the higher angles pass through multiple reflections in their trip through the fiber, and since they actually travel farther than the axial rays, the axial rays reach the termination point of the fiber conductor before the high-angle rays. The resultant time delay in passing through the fiber produces a distortion by spreading the data signal. To reduce this ef-

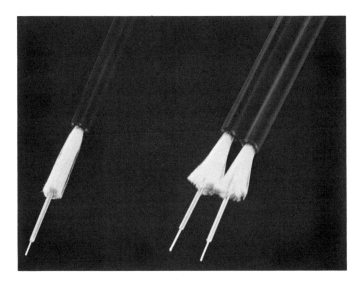

Figure 2.22 Belden Bitlite (TM) fiber optic cables. (Courtesy of Belden Fiber Optics.)

fect, the fiber core can be produced by a chemical vapor deposition (CVD) process which can combine core materials with germanium and boron to produce a core with a gradient index of refraction ranging from 1.47 at the center of the core and decreasing gradually to 1.45 at the cladding interface (see Figure 2.23). In this manner, the angular reflections of high-angle light rays, characteristic of the step-index fiber, is modified to a curvelike reflection which reduces the distance that high-angle rays travel through the fiber and reduces the time delay between high-angle and axial light rays in passing through the fiber. A sample of a 50-μm graded-index optical cable is shown in Figure 2.24.

Graded-index optical fibers are drawn from preforms in the same way as step-index fibers. The major difference is that the glass tube used to form

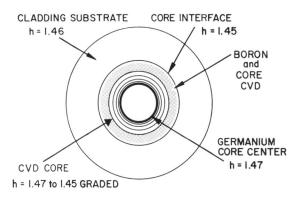

Figure 2.23 CVD graded index optical fiber configuration. (Courtesy of Belden Fiber Optics.)

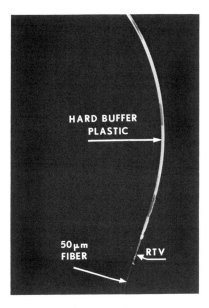

Figure 2.24 ITT Electro Optical [51] Products Div. 50 micron fiber optic cable with photo ID overlay.

the cladding is treated by a CVD process whereby ultrahigh-purity doping and core materials are deposited uniformly in the form of "soot," by means of flame hydrolysis, over the length of the *inner surface of the cladding tube.* After deposition is completed, the submicron particles of "soot" are sintered at high temperature to clear glass and the sintered tube is then collapsed into a solid rod or preform which can now be drawn into optical fibers.

Optical Fiber Parameters

The graded-index fibers are available with a number of different bandwidths, depending on the index profile and core diameter. A one-fiber bandwidth varies from 200 MHz over a distance of 1 km to a bandwidth of over 1000 MHz.

Besides the variations in ray velocities (modal dispersion) another factor which contributes to pulse-spreading distortion is the wavelength intervals over which sources such as LEDs (light-emitting diodes) emit. This causes dispersion because the fiber core refractive index varies with wavelength. Consequently, different wavelengths travel at varying speeds within the fiber, resulting in a pulse spreading termed material dispersion (see Figure 2.25). Because of this property, obtaining maximum bandwidth performance from a fiber must be with a narrow-bandwidth source such as an infrared laser diode that emits over a 2- to 4-nm wavelength interval. LEDs have an emission bandwidth approximately 10 to 20 times that of the laser diode (see Figure 2.26).

Although both modal and material dispersion are limitations on the fiber bandwidth, the magnitude of their effects varies with the type of fiber.

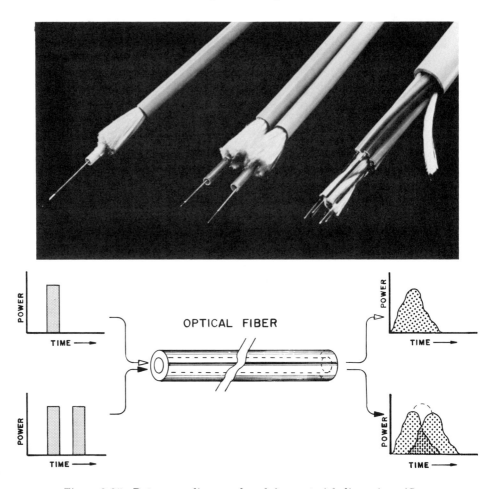

Figure 2.25 Data spreading produced by material dispersion. (Courtesy of Belden Fiber Optics.)

Step-index fibers are limited primarily by the modal dispersion and perform identically for a laser diode or a LED source. On the other hand, graded-index fibers have been designed to eliminate modal dispersion effects and are therefore highly dependent on the source emission wavelength interval. Thus, to utilize their full bandwidth potential requires laser diode source.

In specifying the fiber bandwidth, the frequency at which the signal response falls by 3 dB is typically quoted. This is because the pulse-spreading mechanisms always result in some degree of dispersion. In addition, this response is commonly specified for a narrow-bandwidth emitter such as the laser diode.

Frequency response curves for some 1-km-long step- and graded-index fibers are presented in Figure 2.27. Since bandwidth is length dependent, a

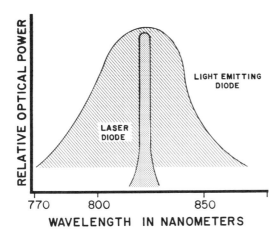

Figure 2.26   Bandwidths for LED diodes and LASER diodes. (Courtesy of Belden Fiber Optics.)

1-km 200-MHz fiber will provide approximately 70-MHz bandwidth at the 3-dB point at a length of 4 km.

The length-dependent bandwidth of optical fibers differs drastically from conventional parallel wire or coaxial cable. For electronic cable the bandwidth varies inversely as the square of the length, while for optical fiber the bandwidth varies inversely only as the 0.75 power of the length. Thus the optical fiber maintains its bandwidth over far longer lengths. A comparison of a 200-MHz-km optical fiber with RG-59 coaxial cable yields the following results:

	−3-dB bandwidth for length:	
	100 m	*1000 m*
Optical fiber	1125 MHz	200 MHz
RG-59 coaxial cable	22 MHz	0.22 MHz

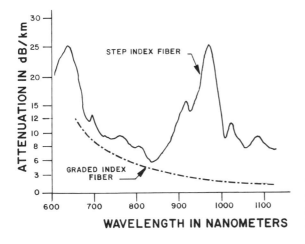

Figure 2.27   Step and graded index fiber attenuation with optical frequencies. (Courtesy of Belden Fiber Optics.)

Attenuation per kilometer is another method of comparing optical fiber cables to coaxial cables. Using the same RG-59 coaxial-cable and graded-index optical fibers at 200 MHz, the attenuation would be as follows:

	Attenuation per length:	
	100 m	1000 m
Graded-index optical fiber	0.3–0.8 dB	3–8 dB
RG-59 coaxial cable	12.5 dB	125 dB

Test measurements of graded-index optical cables simultaneously use dual lambda frequencies of 0.83 μm and 1.3 μm to inject narrow pulses into the fiber to measure attenuation, bandwidth, and dispersion at each frequency in lengths of 1 m and the total spooled length of the draw (1 to 3 km).

CHAPTER REVIEW QUESTIONS

1. Name the four classes of contaminants that produce data pollution. Which has the greatest potential for computer or peripheral circuit damage?
2. Which peripheral is most likely to be damaged by airborne particles?
3. Why are electric motor appliances a problem with computers or their peripherals?
4. What is the cause of ground loops when coaxial cables are used as data links?
5. What is the approximate thickness of the aluminum used for foil-shielded cables?
6. What is meant by *foil optical coverage*?
7. Why is the transfer impedance of braided shields lower than spiral-wrap shields?
8. At what frequency does the combination foil and braid cable's transfer impedance begin to fall?
9. What benefit is obtained by connecting the inner shield, of triax cable, to the center conductor at the source end only?
10. For data transmission in very high EMI/RFI areas, which type of shielded cable would provide the least data pollution?
11. List six situations where optical data cables eliminate problems that shielded cables merely reduce.
12. What is the basic element of fiber optic technology?
13. What is the cause of signal spreading in propagation through an optical fiber cable?
14. Which type of refractive index optical fiber has the higher frequency response?
15. What is the numerical aperture, and how is it derived?
16. What enables a 50-μm optical fiber to be able to withstand elongation test pressure?
17. What are the graded-index fiber's center core index of refraction and cladding interface index?
18. What is modal dispersion?

19. What is material dispersion?
20. How does transmission-line loss of a graded-index fiber cable compare with that of RG-59 coaxial cable over a distance of 100 m?

GLOSSARY

AM. Amplitude modulation: commercial entertainment radio broadcasting in the radio spectrum of 550 to 1600 kHz.

Aperture. The opening that limits the quantity of light that can enter an optical device or instrument; an opening, hole, or slot.

Attenuation. To weaken or reduce in force, intensity, effect, quantity, or value.

Axial. Characterized by or forming an axis, in line with the axis or center.

Cladding. The optical material surrounding the central optical core that produces an index-of-refraction difference that permits total reflection of all light rays entering the acceptance cone of the optical core.

CMOS. Complementary metal-oxide semiconductor: A method by which both p- and n-channel semiconductor devices are produced on the same substrate.

Coaxial. Sharing the same axis; a combination of a signal-carrying, insulated conductor surrounded with a metallic shield of woven-wire braid as a coaxial shielded cable.

Conductor. A metal wire or cable capable of carrying an electrical current; the atom of an element having one to three electrons in its outer shell is considered a good conductor element.

Continuity. Continuous or connected whole without interruption.

Crosstalk. Interference by or reception of other signals from adjacent wires or cables.

CVD. Chemical vapor deposition: a process for producing the central optical core of an optical fiber with an index of refraction graded parabolically from 1.47 to 1.45.

dB. Decibel: a unit of power ratio proportional to the common logarithm of the intensity of two sources; $dB = 10 \, LOG \, (P_{out}/P_{in})$.

Disk drive. A mechanical device with the capability of reading or recording computer data onto or from a plastic disk coated with a ferric oxide magnetic recording medium.

Doping. A chemical process by which an intrinsic material takes on an impurity component that alters the characteristics of the intrinsic material while retaining its basic sturcture; a process by which intrinsic silica or germanium may be doped with trivalent atoms of gallium to produce a P-semiconductor characteristic or with a pentavalent atom or arsenic to produce an N-semiconductor characteristic.

Earth ground. A method of providing an external electrical ground return isolated from an equipment or chassis ground where EMI or RFI interference exists.

EMI. Electromagnetic interference caused by radio or television broadcast signals or by collapsing magnetic fields surrounding current-carrying wires, electric motors, or electric arc welding equipment.

Ethernet. A local-area network for linking computer systems by means of special high-speed data cables.

FCC. Federal Communications Commission: the federal regulatory agency for all forms of radio and television broadcasting in the United States.

Fiber. Herein referred to as the optical fiber or optical waveguide used to transmit data in the form of modulated light frequencies.

FM. Frequency modulation: a method of modulation of a radio carrier frequency in which an audio signal causes the carrier frequency to increase or decrease by the frequency of the applied audio signal.

Foil. A thickness of metal applied to a polyester substrate or tape by the process of vacuum evaporation of the metal. Thicknesses on the order of 10 to 50 μm produce very flexible conductive metallic coatings.

Graded. Contouring a surface or composition of a material; a graded index of refraction is made to vary in index from a high index to a lower index, or vice versa.

Ground loop. A condition where differences in potentials that exist at the points where braid shields are multiply grounded produce unwanted shield currents and inductively couple the shield currents into the signal-carrying conductor.

Impedance. The vectorial sum of the opposition to the flow of alternating current in a circuit containing resistance, inductive reactance, and capacitive reactance.

Infrared. An electromagnetic radiation at the lower end of the visible (red) spectrum of light with a wavelength ranging from 1 mm to 10 μm.

Insulation. A material exhibiting very high resistance to the flow of an electrical current or a current of any frequency in the electromagnetic spectrum; any intrinsic element containing 6 to 8 valence electrons is considered an insulating material.

kHz. Kilohertz: denoting a frequency of 1000 hertz or cycles. The prior term, cycles per second, was renamed hertz to honor the work of Heinrich Rudolf Hertz (1857–1894) in the field of electromagnetic radiation.

Laser diode. A multilayer light-emitting diode capable of emitting a very narrow band of optical frequencies; a seven-layer GaAl GaAsAl semiconductor that exhibits the laser properties of producing near-monofrequency optical energy.

LED. Light-emitting diode. A PN-semiconductor diode capable of emitting a wide range of optical frequencies from the red visible spectrum through the invisible infrared spectrum.

Low-pass filter. A combination of inductors and capacitors configured by connections to offer low impedance to frequencies below a cutoff point and high impedance to the same lower frequencies above the cutoff point.

MHz. Megahertz: denoting a frequency of 1,000,000 hertz.

Micron. One millionth of a meter; 0.000039 in.

Nanosecond. 1×10^{-9} seconds.

Parallel. Two conductors running side by side; eight binary bits transmitted simultaneously over eight parallel wires; a parallel input/output port.

Peripherals. The electronic devices surrounding a computer, such as the CRT monitor, keyboard, disk-drive units, and the printer. Also included as a peripheral are the devices and sensors under the control of the computer.

Permeability. The ease with which a material can accept lines of magnetic force; the ability of a material to concentrate a large number of magnetic lines of force into a small area.

Preform. A fused core-to-cladding boule from which an optical fiber is drawn.

Pristine. Pertaining to the earliest period or state in the fabrication of an optical fiber.

Propagation. To transmit light, heat, or sound through space or a physical medium.

Resistance. The opposition to the flow of electrical direct current; an electrical component with the property of limiting the flow of direct current in an electronic circuit.

RFI. Radio-frequency interference.

RS-232 serial port. A means of transmitting data serially, one bit at a time, one after another, like railroad cars following in line behind the engine.

RTV. Room-temperature vulcanizing: a silicone type of plastic that will cure or vulcanize at room temperature.

Silica. A form of glass composed of melted sand and other elements to assure transparency and strength.

Sintered. To bring about agglomeration by heating.

Spark gap. The first method of transmitting a radio signal—by causing an arc to jump across the spacing between two electrode balls.

Static electricity. A form of electricity generated by friction; an electrical charge created by walking across synthetic fiber carpet; electrical discharges (lightning) from friction of moisture-bearing cloud formations.

TV. Television.

Video. A visual image displayed on a computer display terminal.

V_t. Symbol for total voltage.

White elephant. Any worthless object that no one wants.

Z_t. Symbol for total impedance.

3
Computers and Peripherals

Interfaces advertised as being compatible will be incompatible.
Murphy

The first impression one gets from the word *computer* is an instant vision of a massive mainframe surrounded by dozens of reel-to-reel tape data storage banks. This impression has been falsely created through the media of science-fiction motion pictures and television programs. The facts tell us that any computer can perform external control functions provided that it has the necessary interfaces and programs required to effect control functions. To illustrate this point, we will take one of the popular microcomputers and show the internal interface circuits that would enable the microcomputer to perform control functions.

Figure 3.1 shows the author's Apple computer used for both word processing and experimental control of robots. Its external peripherals are the dual disk-drive units and the CRT monitor. For the computer's main logic board to be able to communicate to the disk-drive units, it must have a parallel interface card installed in one of the eight edgecard connector sockets on the back side of the main logic board and a 20-conductor flat ribbon-type cable linking the computer disk-drive interface to the disk drives. Figure 3.2 shows the dual-disk-drive interface card being inserted in edgecard socket 6 at the rear of the Apple's main logic board. It is important to follow the instructions for installing any optional interface because each make of computer has different assignments for interface locations. Figure 3.3 further demonstrates the installation of a different type of parallel interface. In this case the parallel interface is for an external printer and functions in a differ-

Figure 3.1 Author's Apple used for word processing and experimental robot control.

Figure 3.2 Installing an Apple disk drive card into the accessory sockets.

Figure 3.3 Installing the Apple parallel printer card.

ent way from the disk interface in that the printer does not transmit character data bidirectionally.

COMPUTER INTERFACE LIMITATIONS

In many cases, computers like the Apple become so filled with optional interfaces that there is no space left to add more. Figure 3.4 shows the following interfaces in the author's Apple computer:

A. Additional random access memory (RAM)

B. Communications serial interface

C. Z-card software interface

D. Dual-disk interface card

E. Light-pen interface control card

The left side of Figure 3.5 illustrates the relationship between a computer's central processing unit (CPU) and its supporting internal circuits and interfaces that make up the computer mainframe. The right side of Figure 3.5 shows the external peripherals and sensors under the control of the CPU. The interfaces for serial, parallel, disk drive, and the analog-to-digital and digital-to-analog (A/D-D/A) converter are shown as external bidirectional links by the black input and white output arrows. The computer's CPU connections to the RAM-ROM, keyboard, and all interfaces to external peripherals are by bidirectional paths on the main logic board. There are several

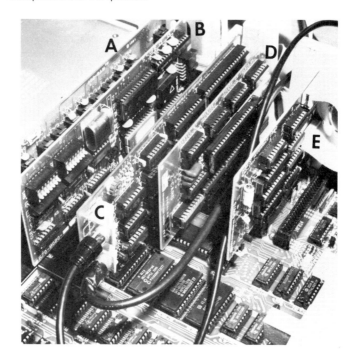

Figure 3.4   Interfaces used in the author's Apple computer.

key circuits that will be described in detail; the CPU and the serial/parallel
I/O ports.

The Central Processing Unit

The central processing unit (CPU) is the only part of a computer where data
can be entered, manipulated, stored, retrieved, and returned to the user. All
of the remaining sections of the computer serve to support the CPU with
data storage locations, control timing, and direction of data movement in
and out of the CPU. The CPU is also known as a microprocessor (MPU),
which is a single integrated circuit containing all the functions for processing
data. The actual size of the microprocessor chip, shown in Figure 3.6, is approximately 9 by 11 mm (0.354 by 0.432 in.), which is mounted in a 40-pin
dual-in-line package (DIP). The electrical connections between the MPU chip
and the DIP connector pins are pressure-welded solid gold wires on the order
of 0.002 in. diameter. The electronic circuitry of the MPU is too small to be
seen by the human eye. An enlarged photomicrograph of the Rockwell International PPS-4/1 microcomputer chip is shown in Figure 3.7, which should
give the student an understanding of the complexity of the MPUs circuit
design.

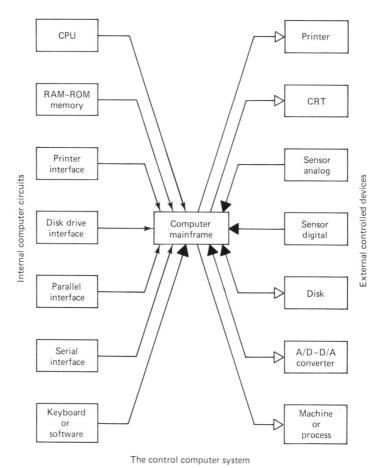

The control computer system

Figure 3.5 Interfaces and control functions of the central processing unit.

The PPS-4/1 One-Chip Microcomputer

The parallel processing system (PPS) of Rockwell International offers 10 production models of the PPS-4/1 family of one-chip microcomputers. They provide a choice of 8-bit ROM from 640 to 2048 bytes and 4-bit RAM from 48 to 128 bytes. The I/O line options provide the design engineer with I/O access to 39 lines. The designer has the options of selecting 28- or 40-pin DIP or 42- to 52-pin quad packaging. A complete chart of the PPS-4/1 microprocessor options is shown in Figure 3.8. To better visualize the on-board functions of the PPS-4/1 Microcomputer, Figure 3.9 shows a block drawing of the systems architecture. The following function descriptions are general in nature and do not include the detail necessary for use as design parameters.

Figure 3.6   Rockwell International PPS-4/1 single chip microcomputer showing actual size and enlarged views of the actual chip. (Courtesy of Dilithium Press.)

Accumulator and Arithmetic-Logic Unit

Approximately in the middle of Figure 3.9, find the blocks identified A, ALU, and C. The primary working register is the accumulator (A), which ties with the arithmetic-logic unit (ALU) and the carry flip-flop (C) to perform either binary or decimal arithmetic. The accumulator can be loaded with constants by appropriate instructions from the ROM or with variable data

Figure 3.7   A photomicrograph of the PPS-4/1 chip design circuitry. (Courtesy of Dilithium Press.)

loaded from or exchanged with the RAM under control of the data register. Parallel or serial data can be loaded into the accumulator or, alternatively, stored in X or S registers in the form of 4-bit words. The contents of the accumulator may be output for control or data transferred through the A buffer, which consists of four latched, open drain circuits, that holds the data until either new data are received or the power is turned off.

Driver and Receiver Circuits A and X

The output of each of the latches in the A buffer, X buffer, and the discrete input/output flip-flops is via open drain drivers. These either drive to the V_{ss} or float. When power is applied, a power-on reset circuit (PO) will cause all the outputs to be set automatically to the float condition and all flip-flops to be reset. Input and output signals may share the same data lines for register (R-I/O 1–8) and discrete I/O lines. The incoming signal may then take the line high to V_{ss} or to the appropriate low logic level. The input command, in effect, samples the logic level on the pin and inputs it appropriately. It should be noted that the A driver and register can input directly to the ac-

ROCKWELL INTERNATIONAL PPS-4/1 ONE-CHIP MICROCOMPUTERS

FEATURES / MODELS	MM75	MM76	MM76C*	MM76E	MM77	MM78	MM76L	MM76EL	MM77L	MM78L
ROM (X8)	640	640	640	1024	1344	2048	640	1024	1536	2048
RAM (X4)	48	48	48	48	96	128	48	48	96	128
TOTAL I/O LINES	22	31	39	31	31	31	31	31	31	31
Cond. Interrupt	1	2	2	2	2	2	2	2	2	2
Parallel Input	4	8	8	8	8	8	8	8	8	8
Bidirectional Parallel	8	8	8	8	8	8	8	8	8	8
Discrete	9	10	10	10	10	10	10	10	10	10
Serial	–	10	10	10	10	10	10	10	10	10
Package (in-line)	28 pin dual	42 pin quad	52 pin quad	42 pin quad	42 pin quad	42 pin quad	40 pin dual	40 pin dual	40 pin dual	40 pin dual
POWER			–15@ 70mw (typical)					–6.5 to –11 V @ 15mw (typical)		

*Two 8-b or one 16-bit presetable up/down register

Figure 3.8   The Rockwell International's family of PPS-4/1 single chip microcomputers. (Courtesy of Dilithium Press.)

cumulator (R-I/O 1–4), whereas the X driver and register must input to the
accumulator through the X register.

S Register and Shift Counter

The shift register is a 4-bit, parallel in/parallel out shift register which is used
as an auxiliary storage register or as a buffer for the simultaneous serial in/
serial out capabilities of the microcomputer. The 4 bits to be serially output
are loaded into register S from the accumulator. The state of the I/O line is
immediately set by the data in the most significant position. When the I/O
serial instruction is executed or an external shift clock input is received, the
4-bit content of the S register is shifted out with the most significant bit
first. The data shift rate is under the control of the program counter and is
typically half the rate of the internal clock frequency. The I/O serial instruction also causes the shift counter to transmit four shift clock signals to the
external system. If the shift register is under the control of an external shift
clock on the signal line, the shift rate can be any value above or below the internal clock frequency. When 4 bits of data are being shifted out on the I/O
line and there are 4 bits of data to be received, the new data can be shifted
into the S register from the I/O line simultaneously. The contents of the S
register can then be shifted to the accumulator for processing or storage.
Simultaneously, new data can be shifted into the S register if more than
4 bits are to be transmitted out.

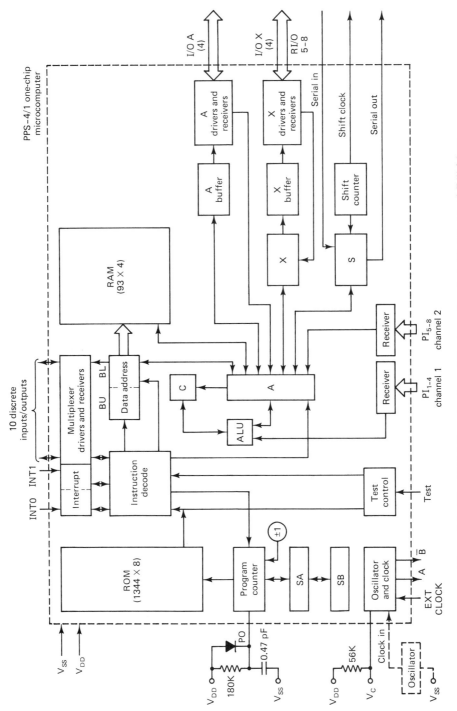

Figure 3.9 The PPS-4/1 system's architecture drawing. (Courtesy of Dilithium Press.)

60

Discrete I/O Ports D-I/O 0 through D-I/O 9

The PPS-4/1 contains 10 discrete I/O lines and two special interrupt input lines. The flip-flops associated with all 12 of these channels may be individually set or reset under program control or automatically reset at the power-up sequence. A buffer flip-flop is associated with each channel selected by the least significant 4 bits of the data address register.

Conditional Interrupts INT-∅ and INT-9

These ports may be addressed by the lower portion of the B register (BL = 11 selects INT-∅ and BL = 10 selects INT-1) in order to set or reset the selected flip-flop. Although these ports are similar to the discrete I/O channels, they do differ in several respects. The Skip On, Selected Low instruction tests the state of the flip-flop, not that of the I/O line. To test the state of the I/O line directly, the INT-0H or the INT-1L instruction for INT-0 and INT-1 permits a pseudo interrupt test capability without preconditioning. The lack of drivers on these I/O lines eliminates the necessity for setting the flip-flops to a predetermined state. A second difference is that INT-0 and INT-1 can be used to detect pulse durations longer than one clock cycle. In this case, the associated flip-flop is preset to the set rate for an incoming positive pulse on INT-0 or for an incoming negative pulse on INT-1. Either signal will reset the flip-flop if its duration is longer than one clock cycle.

Input Channels 1 and 2

Channel 1 is a 4-bit port that automatically adds the input value to the accumulator and then skips the next instruction if no carry is generated. Channel 2 is a 4-bit port which, on command, replaces the contents of the accumulator with the complement of the value on the input lines.

Program Counter

The 11-bit program counter is also referred to as the program register. The contents of the program counter address the ROM to identify the specific instruction to be executed. If no transfer instruction is received, the contents of the program counter are incremented so that the next instruction may be selected. This process repeats until a transfer, or transfer and mark, instruction is executed.

SA Register

When a subroutine call is executed by one of the transfer and mark instructions, the contents of the SA register are pushed into the SB register and replaced by the incremented value of the program counter. When a return instruction is executed, the contents of the SA register are popped into the program counter and the SA register contents are replaced by those in the SB register. The SB register provides a second hardware stack so that two levels of subroutines may be nested in the microcomputer.

Read-Only Memory

The read-only memory (ROM) provides the storage for instructions and constants in the microcomputer. It contains from 640 to 2048 instructional 8-bit bytes depending on the computer type selected. The ROM is under the control of the program counter.

Instruction Decode

Program instructions are decoded in the instruction decode circuits from which control signals are derived and then directed to all appropriate sections of the microcomputer as necessary to perform the programmed instruction or function.

Data Memory

The random access memory (RAM) is used for data memory and consists of ninety-six 4-bit characters. This memory is used to buffer I/O values, hold intermediate results, or as additional registers to be used for timing, counting, comparators, and so on, when the microcomputer is used as a universal logic element.

Clock Control

The PPS-4/1 may be driven by either the internal clock (CLKIN) or an external clock (EXCLK). For either of the clock modes, a 56,000-Ω resistor must be connected between V_C and V_{DD}. The normal internal 80-kHz clock is used to drive the microprocessor in systems where precise timing is a requirement. The external clock may be used to drive the PPS-4/1 at frequencies between 40 and 80 kHz. The external clock P-IN must be tied to V_{DD} and the external clock frequency inputted to EXCLK.

Test

One of the PPS-4/1's advantages is that it is testable at both the factor and user levels. When a test state is indicated by the test line input, the PPS-4/1 goes into a test mode which tests the ROM and allows testing of RAM and instruction logic.

COMPUTER PARALLEL AND SERIAL INTERFACING

Interfacing a computer CPU to external control applications requires the use of parallel or serial I/O ports in addition to the local peripheral interfaces for CRT and disk drives. The following descriptions of parallel and serial I/O interfaces will include the transmitter and receiver options for interfacing the computer to fiber optic data links.

The Peripheral Interface Adapter

There is a third type of parallel interface, called the peripheral interface adapter (PIA), which contains two bidirectional parallel ports in a single 40-pin integrated circuit such as the Motorola MC6820 shown in Figure 3.10. The use of the PIA can be expanded by combining two or more PIAs on a single circuit board. Figure 3.11 illustrates in block diagram form two methods by which the PIA can be used to interface input and output devices to the CPU. A detailed illustration of the PIA block diagram is shown in Figure 3.12. This figure shows the address bus, data bus, and clock connections between the CPU, ROM, RAM, and the PIA. These key connections vary with the type of CPU and its compatible PIA. However, all generally follow the same method for addressing the PIA and differ only in nomenclature and pin connections. Figure 3.13 is a logic circuit diagram to enable you to follow the addressing for the Motorola 6820 PIA.

Figure 3.10   The peripheral interface adapter (PIA) using the Motorola MC6820 chip.

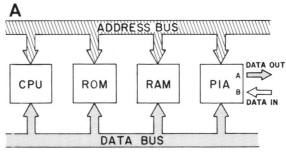

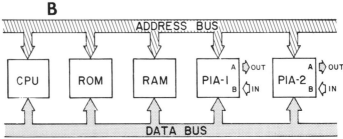

Figure 3.11   Block diagrams of use of the PIA for expansion of parallel I/O ports.

Addressing the Motorola 6280 PIA

Three chip select lines—CS0—pin 22, CS1—pin 24, and $\overline{\text{CS2}}$—are used to select a specific PIA. CS0 and CS1 pins must be high and $\overline{\text{CS2}}$ must be low in order to select the PIA to be used. The three lines connect to address lines A2, A14, and A15. The PIA will be selected by any address when A2 and A14 are high and A15 is low. Thus any hexidecimal address such as 4004, 5004, 6004, or 7004 selects the PIA because the address conditions are met. This is not intended to imply that these are the only addresses that will select a given PIA, for there are literally thousands of combinations in the hexidecimal number system that would meet the address requirements. For this reason, the DMA and A15 are ANDed to ensure that the PIA is selected only if the address is valid. For programming purposes, we must assign four addresses to the PIA, which could be 4004 through 4007, 5004 through 5007, 6004 through 6007, and so on. The final address will be controlled by the DMA line.

The PIA has two register lines, RS$\emptyset$—pin 36 and RS1—pin 35, that connect to the address bus. The RS1 line determines which side of the PIA is to be used. If RS1 is at logic $\emptyset$, side A is used. If RS1 is at logic 1, side B is used. Line RS$\emptyset$ is the register select line. If RS$\emptyset$ is at logic 1, the control register is selected, and if RS$\emptyset$ is at logic $\emptyset$, either the data direction register or the output register is selected, depending on the state of bit 2 of the control reg-

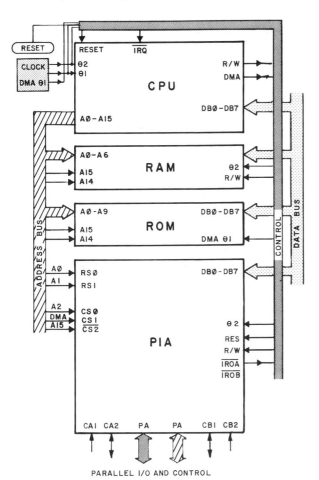

Figure 3.12 PIA address and data bus connections to the computer's CPU, RAM, and ROM.

PARALLEL I/O AND CONTROL

ister. There is one more control line for the PIA, the R/$\overline{\text{W}}$ control at pin 21. If this line is at logic 1, the PIA reads the data on D0 through D7 of the side selected. If this line is at logic 0, the PIA writes the data from the CPU register on the D0 through D7 outputs of the side selected. Figure 3.14 shows a parallel interface containing three PIAs and their control logic circuits. A parallel interface of this type offers a choice of 24 output and 24 input data lines for digital OFF/ON state control and sensor status return.

Serial RS-232-C Interface

The serial RS-232-C interface is a communications interface by which the data on the computer's 8-bit parallel data bus is converted to a bit-in-line or serial form for transmission between the computer and a receiving device (see Figure 3.15). The device that performs the parallel-to-serial conversion is called a universal asynchronous receiver/transmitter (UART). The UART

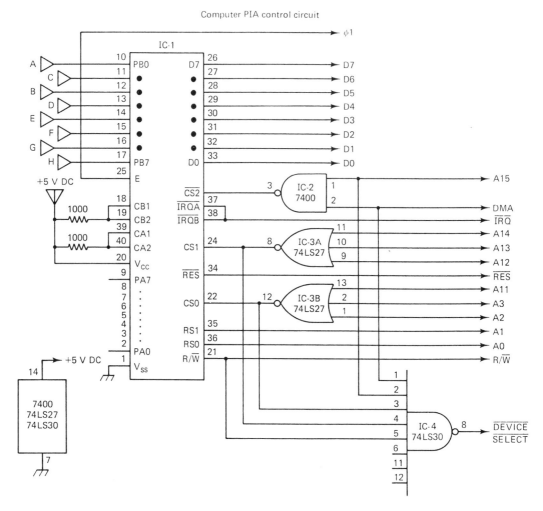

Figure 3.13   Logic diagram of PIA addressing methods.

has the advantage of being capable of transmitting digital data in analog or audio tones for transmission via a modulator-demodulator (modem) interface to telephone lines, or in digital by means of a 20-mA loop for transmission by shielded twisted-pair cables, or for modulating the optical transmitter for transmission via optical fiber cables.

In the analog serial mode, for storing on cassette tape, the data are in the form of audio tones which can be recorded serially onto the tape. Retrieving data stored in this manner requires that the audio tones be converted back to parallel digital data form by the UART. When the RS-232-C serial interface is used in the 20-mA mode, the data is sent to a device called an opto-isolator or opto-coupler, which drives a light-emitting diode (LED)

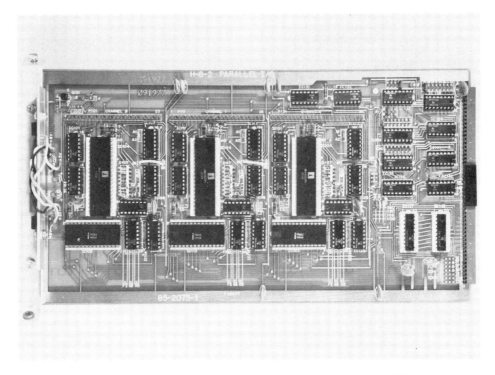

Figure 3.14   Heath/Zenith H-8 parallel interface with three PIA's.

which turns on with a logic 1 and off with a logic 0. The light emitted by the LED is then detected by a phototransistor by effectively increasing its forward bias and causing the loop current to vary in step with the serial data.

Serial Data Transfer Rate

In parallel and serial data transfer, the speed by which an 8-bit byte or word can be transmitted is called *baud*. The RS-232-C EIA standard is 300 baud. However, it is possible, with some serial interfaces, to change the data transfer rate through the use of a four-switch combination to select one of the following transfer rates:

Serial Selectable Baud

110	1200
135	2400
300	4800
600	9600

Each serial interface must be set to the same baud before data can be sent and received between the computer and the device or modem.

Figure 3.15   Heath/Zenith H-8 serial interface with UART.

THE BASIC CONTROLLER BC2

One of the most unique computers designed for industrial control applications exclusively is Action Instruments Co. Inc.'s Basic Controller BC2, shown in Figure 3.16. It has been designed as an industrial machine or process control computer with the capability of serving as an engineering development computer for designing computer-controlled machine installations and automating an industrial process. Figure 3.17 depicts the Basic Controller's world of peripherals, interfaces, and controlled devices. It excels as a slave computer controlling a machine or process under the supervision of a host computer, or it makes an ideal supervisory computer for slave BC2 computers. Add a modem and the BC2 can become a communications computer between a factory and a distant administration office. The bottom line is the BC2's principal function; the job it does best is to control external devices through its 32 sense inputs, 32 flag outputs, and 8 on-board relays. The BC2 design is strictly functional without the cosmetic frills of business computers. In Figure 3.18, you can see that it appears to be nothing more than a circuit-packed motherboard covered with a piece of red plastic. That red plastic cover, shown in Figure 3.19, not only protects the BC2's electronic

Figure 3.16 Action Instruments Co. Inc.'s Basic Controller BC2 computer system. (Courtesy of Action Instruments Co. Inc.)

circuits, but also identifies each edgecard connector, locates LED indicators Ø through 7, and provides LED displays from the least significant bit (LSB) to the most significant bit (MSB) of the data input word. An unusual feature of the Basic Controller is that it does not come with a keyboard, CRT monitor, cabinet, disk or cassette for data storage, power supply, connecting cables, or manual. These are user-selected accessories to eliminate redundancy of items not required when a number of BC2 computers are to be installed as slave computers under the supervision of a host computer or another BC2 computer. In this instance, it would not be necessary to duplicate all of the individual hardware for the BC2 computers serving as slave units.

Although the BC2 is hardly more than an assembled motherboard, do not let its looks deceive you, for behind that red plastic cover is a powerhouse of control functions:

1. Z-80 Industrial Basic Language (ZIBL) Interpreter in ROM (8k)*
2. 16k RAM expandable to 48k RAM and EPROM
3. 4k EPROM (two sockets of 2k each)
4. On-board video circuitry (16 lines, 64 columns)
5. ASCII keyboard interface

*Z80 and ZIBL are registered trademarks of Zilog, Inc.

6. Two RS-232 serial I/O ports
7. Parallel I/O with separate input/output buses
8. 6 vectored hardware interrupts
9. Four groups of single-wire I/O lines expandable externally to 64
   a. 32 sense inputs, 32 flag outputs
   b. 8 liteports with on-board LED status indicators
   c. 8 on-board relays
   d. 8-bit liteport with LED binary data display

THE BASIC CONTROLLER™ WORLD

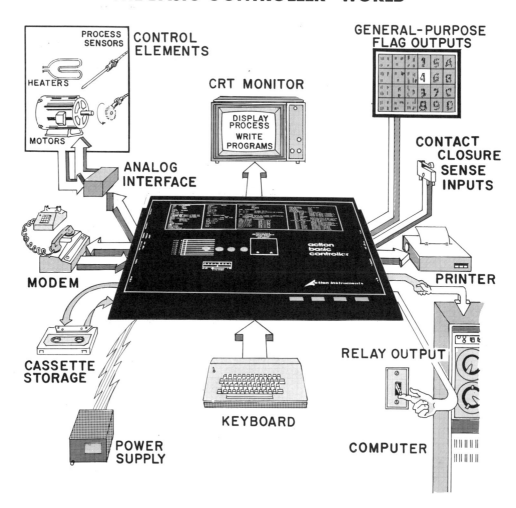

Figure 3.17   Basic Controller's world. (Courtesy of Action Instruments Co. Inc.)

The BC2 sense, Liteport and Relay Features

One of the BC2's most outstanding control features is its 32 sense inputs (see Figure 3.20A). Each sensor is an individually memory-mapped, low-power Schottsky TTL with pull-up resistors to the +5-V dc supply. The logic state of a sensor input is obtained from the LSB of the data word read from the memory address of the sensor input. This configuration permits the sensor input to be connected to a mechanical switch such that when the switch is closed, the sensor input is grounded to indicate a logic 0 and returns to +5 V dc when the switch is opened for a logic 1. A typical ZIBL program for a sensor to be used to indicate if a mechanical switch is open or closed would be:

```
10  DO
20  PR "THE SWITCH IS OPEN"
30  UNTIL SENSOR (5) = 0
40  PR "THE SWITCH IS CLOSED"
```

Running this program will cause the print statement on line 20 to print on

Figure 3.18   BC2's circuitboard. (Courtesy of Action Instruments Co. Inc.)

the CRT monitor "The switch is open" over and over until a switch connected to sensor input 5 changes to a logic 1 state. At this time, the program jumps to line 40 and prints "The switch is closed" on the CRT display.

Figure 3.20B outlines the parallel I/O port at connector J5 and the eight liteports 0 through 7. When the parallel I/O is used for control circuits, the eight liteports indicate the logic states on each of the data lines D0 to D7 at the output of the parallel port.

Liteport	Parallel data line	Hex address
0	D0	FEC0
1	D1	FEC1
2	D2	FEC2
3	D3	FEC3
4	D4	FEC4
5	D5	FEC5
6	D6	FEC6
7	D7	FEC7

To illustrate the liteport/parallel port as a control status indicating port, we can tie in the BC2's 24-hour on-board clock as a real-time program function. A sample program would be written in this manner:

```
10  OPEN %10
15  TIME = #083000
```

LINE 10: OPEN %10 is used to enter current time and write the current clock setting on the supervisory line on the CRT.

LINE 15: TIME = # HHMMSS sets the time to the value after the # character. The clock is automatically set to zero at the power-up start and must be reset in order to have current real time or can be started from zero time to indicate total operating time from power-up.

Having set the current time in the 24-hour format, we can now use the clock as an interval timer in our program.

```
20  IF T = #090000 THEN TURNON LITEPORT 0
30  DO UNTIL T = #130000 THEN TURNOFF LITEPORT 0
40  IF LITEPORT 0 IS OFF THEN TURNON LITEPORT 7
45  DO UNTIL TIME = #131000 THEN TURNOFF LITEPORT 7
50  END
```

On entering the run command, the program will be halted until the on-board clock reaches 090000; then the program will turn on both liteport 0 and parallel port data 0 by placing a logic 1 at each point. This state will be held until 1300 (1:00 P.M.) when both liteport 0 and parallel data D0 will be turned off by dropping the logic 1 to a logic 0. At the same time, a logic 1 will be put on liteport 7 and D7 for 10 minutes or until the clock shows 131000, at which time liteport 7 and D7 will be turned off by changing the logic 1 to a logic 0. Each time a liteport or the parallel port is addressed, the contents of the liteport/parallel output line are displayed as a binary number with LEDs starting with the MSB and going to the LSB, just below the liteport LED display. This provides a visual display of the logic states on each data line of the two outputs of the parallel liteport and parallel port drivers.

The liteports and parallel I/O are not the only sections of the BC2 that can employ the on-board clock for timing functions. Figure 3.20C outlines the eight on-board relays that can be direct driver control relays or where the control voltage and current exceed the ratings of the on-board relays, control, and intermediary "buffer" relay whose primary coil meets the ratings of

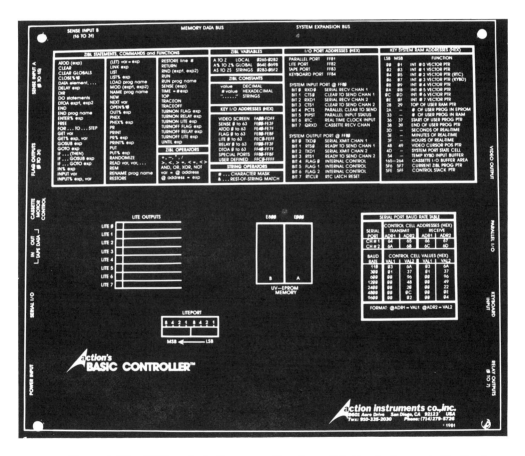

Figure 3.19   BC2's red plastic cover. (Courtesy of Action Instruments Co. Inc.)

Figure 3.20 BC2's sense, liteport, and relay location identifier. (Courtesy of Action Instruments Co. Inc.)

Figure 3.21 BC2's ZIBL and function ROM chips. (Courtesy of Action Instruments Co. Inc.)

the on-board relays. All on-board relays are memory mapped and have maximum current-rated contacts as follows:

Relay number	Maximum contact rating	Duty type	Hex address
0	3 A at 120 V	Heavy	FF00
1	3 A at 120 V	Heavy	FF01
2	3 A at 120 V	Heavy	FF02
3	3 A at 120 V	Heavy	FF03
4	10 VA	Reed	FF04
5	10 VA	Reed	FF05
6	10 VA	Reed	FF06
7	10 VA	Reed	FF07

Figure 3.20D outlines the BC2's RAM, ROM, and EPROM memory area. A minimum of 4k RAM is required to operate the BC2 or may be expanded to 16k on-board RAM. In addition the 16k is also expandable to 48k RAM through bus connections to an off-board RAM expansion card. The ZIBL language and BC2 functional instructions are contained in the 8k ROM chips shown in Figure 3.21. There are two EPROM sockets, identified as BC10, to the right of the ROM chip which can add additional EPROM memory or be used to provide still another feature—producing an EPROM from either program data or copy from other EPROMs.

ZILOG INDUSTRIAL BASIC LANGUAGE

Sample control programs have been given in the functions of liteports and relays in which ZIBL was the program language. To a large extent, many ZIBL statements are similar or identical to BASIC (Beginner's All-purpose Symbolic Instruction Code) language. However, any other comparisons between BASIC and ZIBL would be valueless. ZIBL is primarily a high-speed language requiring only 1 μs to execute a program instruction. It was developed for use with the Zilog Z-80 microprocessor and the Basic Controller BC2.

Zilog Basic Industrial Language Statements

FOR   NEXT
DO   UNTIL
IF   THEN
GOTO
GOSUB   RETURN
LET VAR = exp
LINK exp
DTOA exp,exp

TURNON LITE, RELAY, OR FLAG exp
TURNOFF LITE, RELAY, OR FLAG exp
READ   DATA   RESTORE
DELAY exp
PRINT   PRINT%exp
PR   PR%exp
LIST   LIST%exp
NEW
CLEAR
CLEAR GLOBALS
RUN
RUN NAME
NAME NAME
RENAME NAME
INPUT   LOAD
END
END NAME
STAT = exp
TRACE ON
TRACE OFF
TIME = exp
TRACE ON
TRACE OFF
TIME = exp
IN%exp
OUT%exp

ZIBL Operators

AND, OR, NOT, XOR, +, −, /, *, =,
>=, <=, <>, >, <

ZIBL Functions

MOD(exp,exp)
RND(exp,exp)
ATOD(exp)
SENSE(exp)
TOP
STAT
FREE(o)

ZIBL Constants

#	Hex
	Decimal
"..."	Strings

ZIBL Strings

= character mask in string comparisons
& = remainder of string mask in string comparisons

ZIBL Special Addresses

Video screen	FA00 to FDFF
Sense 0 to 63	FE00 to FE3F
ATOD 0 to 63	FE40 to FE7F
Flag 0 to 63	FE80 to FEBF
Lite 0 to 63	FEC0 to FEFF
Relay 0 to 63	FF00 to FF3F
DTOA 0 to 63	FF40 to FF7F
Ports 0 to 63	FF80 to FFBF
User defined	FFC0 to FFFF
Parallel port	FF81
Liteport	FF82
Tape port	FF83
Keyboard port	FF84

System Input Port at FF80H Bits

Bit0	RDX0	Serial receive channel 0
Bit1	CTS0	Clear to send channel 0
Bit2	RXD1	Serial receive channel 1
Bit3	CTS1	Clear to send channel 1
Bit4	Unused	
Bit5	PIPST	Parallel input port status
Bit6	RTC	Real-time clock as input
Bit7	QRXD	Cassette receive channel

System Output Port at FF80 Bits

Bit0	TXD0	Serial send channel 0
Bit1	RTS0	Ready to send channel 0
Bit2	TXD1	Serial send channel 1

Bit3	RTS1	Ready to send channel 1
Bit4	FLAG0	Internal control
Bit5	FLAG1	Internal control
Bit6	FLAG2	Internal control
Bit7	RTCLR	Real-time clock clear

The ZIBL programming language employed by Action Instruments Co., Inc.'s Basic Controller BC2, was designed to enable anyone with a working knowledge of BASIC to master control programming with a minimum of learning time. In general, the ZIBL language includes a number of BASIC commands and functions which perform in a similar manner in ZIBL but with the extra advantage of the control of memory-mapped functions.

CHAPTER REVIEW QUESTIONS

1. Name three sections of a computer's mainframe.

2. Which would be considered a computer peripheral: a telephone or a modem?

3. How many conductors should a cable link between the computer and a disk-drive unit have?

4. What are the approximate dimensions of a microprocessor circuit chip?

5. What does PPS stand for?

6. Dual-in-line integrated-circuit packages are made in 6-, 8-, 14-, 16-, 20-, 24-, 28-, and 40-pin configurations. How many pins would the DIP have if it were a microprocessor?

7. How many bits does the PPS-4/1 program counter handle?

8. Name two kinds of memory that are considered read-only forms of memory.

9. Name three types of interfaces for interfacing a computer data and address bus to external circuits.

10. In ZIBL, what is necessary to program the BC2 to turn on a relay at 2:00 P.M., then turn it off an hour later?

11. How long does a Z-80 microprocessor require to execute a program instruction?

12. What is the difference between a host computer and a slave computer?

13. What symbol, in ZIBL, denotes hexidecimal?

14. What would a mechanical switch sensor input to a BC2 computer: decimal, hexidecimal, or binary data?

15. In BASIC language a FOR statement must be followed by a NEXT statement. Does this also apply to ZIBL?

16. Correct this program if you find it in error:

```
10  IF SENSE (3)=0 THEN PRINT "THE SWITCH IS ON"
20  DO UNTIL SENSE (3)=1 THEN PRINT "THE SWITCH IS OFF"
```

GLOSSARY

AC. Auxiliary carry.

Accumulator. A general-purpose storage register serving as a source and destination register for manipulated data.

Address. The numerical location of a data storage location in memory or in a temporary register.

ALU. Arithmetic-logic unit.

Ampere. The unit of current.

Architecture. The design configuration of a microprocessor.

Arithmetic-logic unit. The section of a microprocessor where mathematical and logic operations are performed on data received from the data input and program counter.

Basic. Beginner's All-purpose Symbolic Instruction Code: a language for computer programming.

Baud. Denoting the speed, in bits per second, at which data can be transmitted via a serial I/O port.

Binary. A number system to the base 2; the numerical form of data entered, stored, or retrieved from memory.

Bit. One digit of a binary number.

Buffer. A circuit that isolates one circuit from a driving circuit; an interface between one circuit and another.

Byte. An 8-bit binary number or word.

C. Carry.

Card. A printed circuit board containing the electronic components and circuitry that can be inserted into a computer's accessory sockets to perform the function of interfacing the computer's address and data buses to an external device.

Central processing unit. A microprocessor; a section of a computer capable of receiving input data, manipulating the data, and moving the data into and out of storage locations during and after processing and returning them to use as a display on a CRT monitor or printed copy.

Chip. A small silicon wafer on which the circuits of a microminiature semiconductor are produced; a wafer containing all the circuits of a microprocessor or logic gate.

CLKIN. Clock input.

Clock. A device capable of producing a frequency or a fixed number of pulses per second which serves as a timing means to control the functions of a CPU or microprocessor.

CMOS. Complementary metal-oxide semiconductor.

CPU. Central processing unit.

Data. A general term denoting all numbers, characters, or symbols that can be processed or manipulated by a computer.

DIP. Dual-in-line package.

Edgecard. A printed circuit board having fingerlike appendages on one end for the purpose of making electrical contact when inserted into a compatible socket.

EPROM. Erasable programmable read-only memory.

EXCLK. External clock.

Flip-flop. An electronic circuit or device having two stable states which may be shifted from one state to another by a control signal from an external source.

Float. An unstable state of an electronic circuit or device that can be changed to a stable state by a control signal from an external device.

Instruction code. A 7-bit code by which the instruction register can identify 256 microcomputer functions.

I/O. Input/output.

Liteport. An addressable parallel port that indicates the logic value on the addressed data line by a LED, where a logic 1 turns on the LED and a logic 0 turns off the LED.

Mainframe. Refers to the main parts of a computer: the microprocessor, memory components, associated interfaces, and power supply.

Parallel. A method by which an 8-bit byte or word is transmitted over eight side-by-side lines or paths simultaneously.

Peripheral. The internal and external circuits or devices served by the central processing unit or microprocessor.

Peripheral interface adapter. A parallel buffer device between the data bus of a computer and an external parallel-served device.

PPS. Parallel processing system.

RAM. Random access memory.

Real time. An internal computer clock programmed to indicate the local time in hours, minutes, and seconds.

Register. A temporary storage location within the central processing unit or microprocessor.

ROM. Read-only memory.

RS-232-C. The designation for a serial interface capable of transmitting 300 8-bit bytes or words in 1 second.

Serial. A method by which an 8-bit byte or word is transmitted one bit at a time over a single conductor with a shielded ground return.

TTL. Transistor-to-transistor logic.

ZIBL. Zilog Industrial Basic Language: a high-speed instruction execution code for Z-80 microprocessors.

4

Principles of Data Acquisition and Conversion

Data acquisition and conversion systems are used to process analog signals and to convert them to digital for subsequent processing or analysis by computer or for data transmission. In general a *transducer* takes a physical *parameter* such as pressure, temperature, strain, or position and converts it to an electrical voltage or current. Once in this form, all further processing of the signal is done by electronic circuits. After the *analog* processing is complete, the signal is converted to digital form by an *analog to digital converter* and then fed to a variety of possible digital systems such as a computer, digital controller, digital data transmitter, or digital data logger.

A complete representative data acquisition and conversion system is illustrated in Figure 4.1. This diagram shows the various components required for an interconnected system. The input to the system, the physical parameter to be measured, is converted to electrical form by the transducer and then fed to an *amplifier*. The function of the amplifier is to the high level (1 to 10 volt) signal which is necessary for further processing. The signal from the transducer may be a millivolt level signal, a high source impedance signal, a differential signal with common mode noise, or a current signal. In any of these cases, the amplifier is used to convert the signal to a high-level

*This chapter is reprinted from *Microcomputers for External Control Devices* by James A. Gupton, Jr. © 1980 by Dilithium Press. *Author's note:* The conversion of analog signals to digital is such an important factor in the use of computers for the control of any device requiring variable or analog control and analog sensor status feedback that permission to reprint this chapter, from copyrighted publications, has been granted by Datel Systems, Inc. and Dilithium Press.

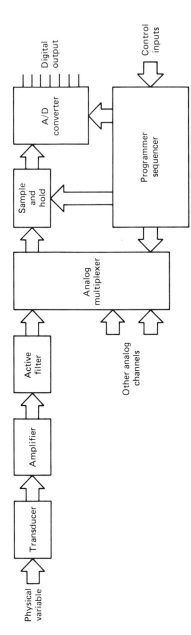

Figure 4.1 Complete data acquisition and conversion system.

voltage which can be used to drive the next analog circuit. An operational amplifier or instrument amplifier is used to accomplish this. The amplifier is followed by a *low pass active filter* which is used to eliminate high *frequency* components or noise from the signal.

There may also be a need to perform some *nonlinear* operation on the signal such as squaring, linearizing, or multiplication by another function. Such operations, which may be performed by an analog multiplier or other nonlinear circuit, also require high level signals to maintain good accuracy and may be performed either before or after the active filter.

The signal then goes to an analog *multiplexer* which performs a time division multiplexing operation among a number of different signal inputs. Each input channel is *sequentially* connected to the output of the multiplexer for some specified period of time. The circuits which follow the multiplexer are thus time-shared among a number of analog signals. The output of the analog multiplexer goes to a *sample and hold* circuit which samples the output of the multiplexer at a specified time and then holds the voltage level at this output until the analog to digital converter performs its conversion operation. The timing and control of this system is accomplished by a programmer-sequencer circuit which controls the multiplexer, sample and hold, and A/D converter. The programmer-sequencer in turn is controlled by digital control inputs from a data processor.

QUANTIZING THEORY

The operation of quantizing a signal is illustrated by the quantizer transfer function shown in Figure 4.2. *Quantization* is the process of converting a continuous analog input into a set of discrete output levels. The analog input is shown on the horizontal axis and the discrete levels on the vertical axis. The discrete output levels can be identified by a set of numbers such as a binary code. The two processes of quantization and coding represent the basic operation of analog to digital conversion.

The quantizer transfer function has a number of important characteristics. The function shown is ideal, with analog decision levels at values of 0.5, 1.5, 2.5, etc. The decision levels are set at values that bracket true levels. In other words, an analog input value of 1 should correspond to a binary output level of 001. The analog 1 value is halfway between the decision levels of 0.5 and 1.5. Thus an analog value of 1 ± 0.5 is read out as a digital 001. The distance between decision levels is Q, the quantization size or bit size. A quantizer with a binary output code has $2n$ discrete output levels with $2n - 1$ analog decision levels. The decision levels in an actual quantizer would not be precise, but would have a finite uncertainty band around them. For an analog value within this uncertainty band, the output could be at either of the two discrete output levels. In addition, in an actual quantizer,

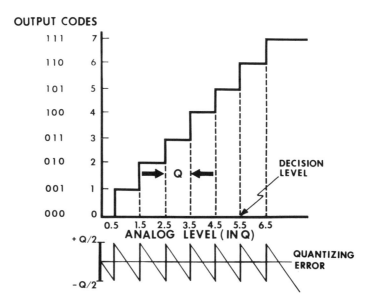

Figure 4.2   Transfer function for quantizer.

the decision levels would not necessarily be precisely at the correct analog values, but would miss these values to some degree due to *nonlinearity*, *offset*, and *gain errors*.

If the input to the quantizer is moved linearly through its full range of values and subtracted from the discrete output levels, an error signal will result. This error is called "quantizing error" and is an *irreducible* error due to the quantizing process and is dependent on a number of quantizing levels, or resolution, of the quantizer. When the quantizing error is plotted, as shown in Figure 4.2, it has the form of a *sawtooth* waveform with a peak to peak value of Q. The output of the quantizer can be thought of as the input analog signal with quantization noise added to it.

Thus the output, which is restricted to a finite number of discrete values, jumps from one value to the next as the input moves through its full range. The quantization error is zero only at the points midway between the decision levels. The peak value of quantization noise is $Q/2$ and the RMS value can be computed from the triangular shape, and it is found to be $Q/2$ × 1.732. Although the quantization noise can be reduced by increasing the resolution of the quantizer, there always remains a quantization uncertainty of at least ± $Q/2$ for any quantizer.

An A/D converter performs the operations of quantizing and coding a signal in some finite amount of time. The time to do this depends upon the converter. The speed of conversion required in a particular situation depends on the time variation of the signal to be converted and the amount of resolu-

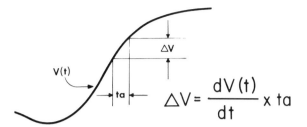

ta   **Aperture time**   Figure 4.3   Aperture time and
ΔV   **Amplitude uncertainty**   amplitude uncertainty.

tion required. The time to make a measurement or conversion is generally called the *aperture time.*

Aperture time can be considered to be either a time uncertainty or an amplitude uncertainty. As shown in Figure 4.3, the aperture time and amplitude uncertainty are related by the time rate of change of the signal. For the particular case of a sinusoidal signal being converted, the maximum rate of change occurs at the zero crossing of the waveform and the amplitude change is:

$$\Delta D = \frac{d}{dt}(V \sin \omega t)_{t=0} \times t_a = V t_a$$

giving

$$\frac{\Delta V}{V} = t_a = 2\pi f t_a$$

From the result we can determine, for example, the aperture time (or conversion time) required to digitize a 1 kilohertz signal to 10 bits resolution. This is a resolution of 1 part in 210 or 0.1% and using this equation:

$$t = \frac{\Delta V}{V} \times \frac{1}{2\pi f t} = \frac{0.001}{6.28 \times 10^3} = 160 \times 10^{-9}$$

The result is a required aperture time on only 160 *nanoseconds* to remain within 1 bit (0.1%) of resolution due to the rate of change of the signal. It can be seen from this result that to convert even a slowly varying signal to moderate resolution levels requires an extremely fast and therefore, expensive analog to digital converter. Fortunately, there is a simple and inexpensive way around this problem by the use of the sample and hold circuit which can reduce the aperture time considerably by taking a rapid sample of the signal and then holding its value for the required conversion time. The aperture time required for sinusoids of other frequencies and different resolutions is summarized by the graphs shown in Figure 4.4.

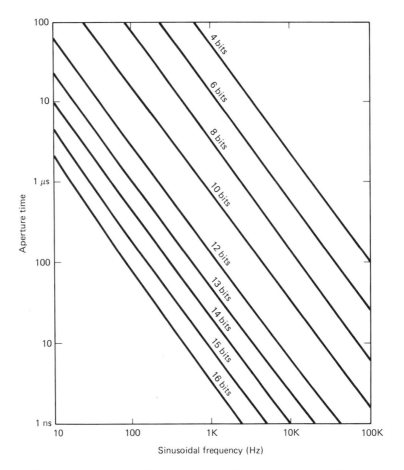

Figure 4.4   Aperture time for a given frequency and resolution.

SAMPLING THEORY

The operation of sampling is illustrated in Figure 4.5 which shows an analog signal and a *train* of periodic sampling *pulses*. The pulses represent a fast-acting switch which connects to the analog signal for a very short period of time (i.e., closes), and then opens for the remainder of the period. Sampling pulses thus have a very short ON time compared to the total period. The result of the sampling process is identical with multiplying the analog signal with a train of pulses of *unity* amplitude.

The resultant *modulated signal* is shown in Figure 4.5c, where the amplitude of the analog signal is preserved in the *modulation envelope* of the pulses. If the *switch-type sampler* is replaced by a *switch* and a *capacitor*, then the analog signal is sampled and stored until the next sample pulse, with

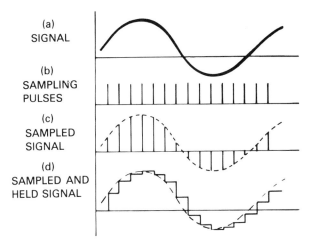

(a)
SIGNAL

(b)
SAMPLING
PULSES

(c)
SAMPLED
SIGNAL

(d)
SAMPLED AND
HELD SIGNAL

Figure 4.5 Signal sampling
process.

the result shown in Figure 4.5d. This type of sampler is called a sample and
hold circuit.

The purpose of signal sampling is the efficient use of data processing
equipment or data transmission facilities. A single *data transmission link* can
be used to transmit many channels of information by simply sampling each
channel periodically. Likewise, for the efficient use of data processing equipment to *monitor* and control a process, for example, it may only be necessary
to sample the state of a process once every 5 minutes, perform a computa-
tion and corrections, and then free the computer during the remaining time
for other task. Continuous monitoring of a single information channel by a
computer would be very expensive indeed. Since the A/D converter may be
the most expensive component in the data conversion system, it is more economical to use a single A/D converter to sample a number of information
channels.

An important and fundamental question to ask about sampled-data
channels is "How often must I sample a given signal in order not to lose information from the signal?" It seems obvious that all useful information can
be extracted by sampling a slowly changing signal at a rate much faster than
any change occurring, and similarly, if a signal is significantly changing value
between samples, information is being lost. The answer is contained in the
well-known sampling theorem which can be stated as follows:

> If a continuous bandwidth-limited signal contains no frequency compo-
> nents higher than f_c, then the original signal can be completely recov-
> ered without distortion if it is sampled at a rate of at least $2f_c$ samples
> per second.

The sampling theorem can be illustrated by the frequency spectra
shown in Figure 4.6. Figure 4.6a shows the spectrum of a continuous signal with frequency components limited to frequency f_c. When this signal is

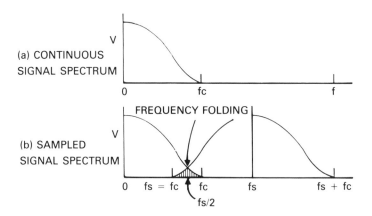

Figure 4.6   Spectra showing frequency folding.

sampled at a rate f_s, the modulation process results in the signal spectrum shown in Figure 4.6b. Here, because the sampling rate is not sufficient, some of the high frequency components of the signal are folded back onto the signal spectrum. This effect is called *frequency folding*. In the process or recovering the original signal, the folded frequency components cause *distortion* since they cannot be separated or distinguished from the original signal. It can be seen from Figure 4.6 that by changing the sampling rate such that $f_s - f_c > f_c$, we obtain the results that $f_s > 2f_c$ which demonstrates the sampling theorem. Frequency folding is eliminated either by using a high enough sampling frequency or by filtering the original frequency to eliminate any frequency components above one half the sampling rate. It should be noted, however, that in practice there is always some frequency folding due to wideband noise and nonideal filters and that one must attempt to reduce the effect to negligible proportions.

The effect of sampling at too low a frequency is called *aliasing* error. Figure 4.7 illustrates this by showing a periodic signal which is sampled at a rate less than twice per cycle. The sample amplitudes are shown connected by a dotted line which obviously has a period quite different from that of the original signal and therefore is an alias. From Figure 4.7 it can be readily seen that if the waveform is sampled at least twice per period as required by the sampling theorem, its original frequency is preserved.

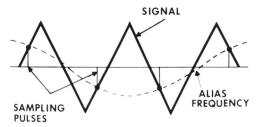

Figure 4.7   Alias frequency caused by inadequate sampling rate.

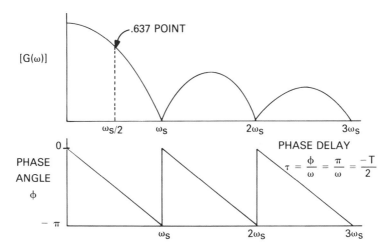

Figure 4.8 Sample and hold transfer.

The sample and hold device, which is so commonly used in data conversions systems and also in analog multiplexed data systems, has some important characteristics which should be briefly discussed. An ideal sample and hold or zero order hold as it is also known, takes a sample in zero time and then holds the value of the sample indefinitely with perfect accuracy. In practical units, a sample is taken in a time period which is short compared to the holding time. During the holding time there is some change in the output which is small compared to the system accuracy. The effect of this process on a continuous analog input signal can be determined by finding the transfer function of a sample and hold. The transfer function of the device is found to be:

$$G(j\omega) = \frac{1 - e^{-j\omega T}}{j\omega} = \frac{2\pi}{\omega_s} \frac{\sin \pi(\omega/\omega_s)}{\pi(\omega/\omega_s)} e^{-j\pi} (\omega/\omega_s)$$

where T is the sampling period and ω_s is the sampling radian frequency. The magnitude and phase of this function are plotted in Figure 4.8 which shows that a sample and hold device acts like a low pass filter with a cutoff frequency of approximately $f_s/2$ and a phase delay of $T/2$ or one half of the sampling period. Circuit characteristics of the sample and hold will be discussed in the following section.

AMPLIFIERS AND FILTERS

The first part of a data acquisition and conversion system is concerned with extracting the signal which is to be measured. The initial processing of the signal is done with an amplifier, a *filter* and possibly a *nonlinear operator*.

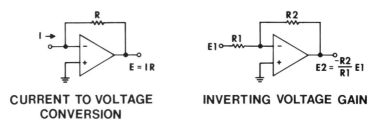

CURRENT TO VOLTAGE CONVERSION   **INVERTING VOLTAGE GAIN**

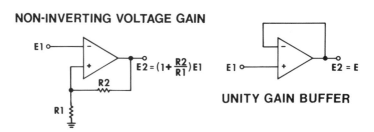

Figure 4.9   Operational amplifier configurations.

The purpose of the amplifier is to perform one or more of the following functions: boost the *amplitude* of the signal, *buffer* the signal, convert a *signal current* into a voltage or separate a differential signal from common mode noise. For most data conversion systems, the desired voltage level of the amplifier is 5 or 10 volts full scale. This is the level accepted by most analog multiplexers, sample and holds, and A/D converters to give the best accuracy. *Operational* or *instrument* amplifiers are usually used to perform the signal translations described previously. Some of the operational amplifier configurations used are shown in Figure 4.9 with their output relationships. In general, an operational amplifier is a good choice for single-ended signal inputs. Although it can be used in some cases for differential signal inputs, the instrument amplifier or data amplifier is the best choice. This type of amplifier is a closed loop configuration as shown in Figure 4.10. It has the following characteristics:

1. *High impedance differentical* inputs
2. Wide range of gains set by a single external *resistor*
3. High common mode rejection ratio

The "common mode rejection ratio" is an important parameter of differential amplifiers. An ideal differential input amplifier would respond only to *voltage* differences between its input terminals without regard to the voltage level common to both inputs. In practice, however, there is a variation in the balance of the differential amplifier due to common mode voltage which results in an output even when the differential input is zero. The common mode error voltage is a nonlinear function of the input common mode voltage.

Definition. Common mode rejection ratio is the ratio of the common mode voltage to the common mode error referred to the input and is generally expressed in dB

$$\mathrm{CMRR} = 20 \log 10 \, \frac{V_{cm}}{e_{cm}}$$

where V_{cm} is the common mode voltage and e_{cm} is the common mode error referred to the input. CMRR is a function of both common mode voltage and frequency. At even moderate frequencies, a high CMRR can be significantly degraded by small unbalances in source impedance and input capacitances.

Following the amplifier in our system, it may be necessary to use a *low pass filter*. Filtering is used in a data acquisition system for two reasons: to limit the bandwidth of the processed signal to less than half the sampling frequency in order to eliminate frequency folding and to reduce either manmade or electrically generated noise in the system. Man-made noise usually has some identifiable characteristic such as periodicity and regular shape and can be eliminated by some specific technique such as a notch filter. Thermal or Johnson noise, on the other hand, is minimized by restricting the *bandwidth* of a system to the minimum required to pass the necessary signal components.

No filter does a perfect job of eliminating noise or undesirable fre-

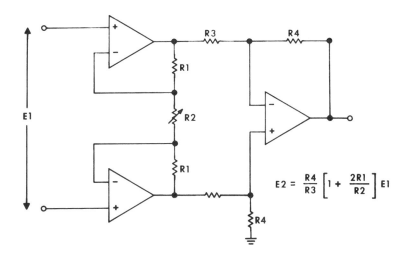

Figure 4.10 Instrument amplifier configurations.

quency components and, therefore, the choice of a filter is always a compromise. Ideal filters with flat response, infinite cutoff *attenuation* and linear response are simply mathematical filters, and not physically realizable. In practice the engineer has the choice of a cutoff frequency, an attenuation rate, and phase response based on the number of poles and filter characteristics chosen. The effect of overshoot and nonuniform phase delay must also be considered.

Active filters are very popular due to the number of excellent features giving them an advantage over the older RLC type passive filters. They eliminate *inductors* and the associated *saturation* and temperature stability problems. The response of an active filter can be accurately set by temperature-stable capacitors and resistors. In addition, they overcome the problems of *insertion loss* and *loading effects* by their use of operational amplifiers. They are, however, limited in frequency response.

SETTLING TIME

A parameter which occurs very often in data acquisition and conversion systems is "settling time." Settling time is defined as the time elapsed from the application of a full-scale step input to an amplifier to when the output has entered and remained within a specified band around its final value. Although this definition is stated in terms of an amplifier response, settling time is also a specification of other components such as D/A converters, analog multiplexers, and sample and holds. It has essentially the same meaning in these other cases although in some of them the input step is actually applied by turning on a switch and, therefore, a switch time is included in the

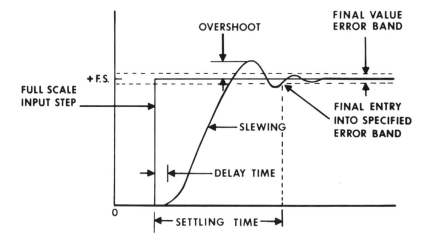

Figure 4.11 Amplifier slew time.

settling time. In case of sample and holds, the equivalent of settling time is the *acquisition time* and includes both the turn on time of the sampling switch and the charging time of the holding capacitor to its final value.

The *settling time* of a fast amplifier is illustrated in Figure 4.11. After the application of the input step, there is a small delay time after which the amplifier begins to *slew* or change its output at the maximum possible rate. During this time the amplifier is in a nonlinear or *saturated state*. The output then *overshoots* its final value, recovers from saturation, and finally settles into its specified error band after a small amount of ringing. The definition specifies that the amplifier must enter and remain in the error band rather than just enter it the first time.

Settling time, unfortunately, is not readily predictable from other amplifier parameters such as bandwidth, slew rate, or overload recovery time although it depends on all of these. It is also dependent on the amplifier's open loop response characteristic, *dielectric* absorption by internal capacitors, output load capacitance, and input capacitance. An amplifier must be designed and optimized for settling time, and settling time is a parameter which must be evaluated by testing.

In addition to wide bandwidth, fast slew rate, and fast overload recovery, another important requirement of a fast settling amplifier is that it have a true single pole open loop response characteristic. This means a smooth 6 dB per *octave* frequency roll-off characteristic. Such an amplifier can never settle to its final value in less time than that derived from the number of time constants required to reach this accuracy. Figure 4.12 shows a plot of the output error as a function of the number of time constants. The settling time may actually be quite a bit larger due to slew rate limitations and other

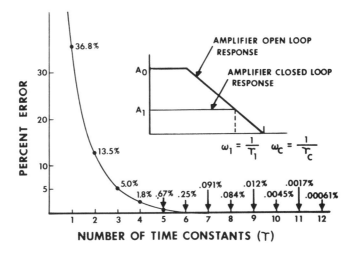

Figure 4.12 Error as a function of the number of time constants for a single pole response.

problems. An amplifier with a closed loop bandwidth of 1 MHz has a time constant of 160 nanoseconds and a settling time within 0.01% of final value requires at least 9 time constants of 1.44 microseconds.

If an amplifier does not have a single pole response and, therefore, an uneven gain roll-off characteristic, its output response may get to the vicinity of the error band quickly but then require a very long time to actually enter it. Similarly, it may overshoot the error band and take a long time to enter and remain inside it. This is the case for amplifiers with pole-zero mismatches in their response characteristics.

Modern, fast settling amplifiers are generally specified to 0.1% or 0.01% settling for small closed loop gains and usually have settling times of less than 1 microsecond. In data acquisition systems, these amplifiers are useful in the design of sample and hold circuits, buffer amplifiers for analog multiplexers or A/D converters, and output amplifiers for fast D/A converters. In many cases, fast settling amplifiers are required even though slowly changing signals are being processed. This is because of the high data switching rate through the analog multiplexer.

DIGITAL CODING

A/D and D/A converters relate analog and digital values by means of an appropriate *digital code*. The codes used are various binary related codes, the most common of which is the natural binary. A binary number is represented as

$$N = a_n 2^n + a_{n-1} 2^{n-1} + \ldots + a_2 2^2 + a_1 2^1 + a_0 2^0$$

where the coefficients a_n assume the values of "0" or "1." Table 4.1 shows the decimal equivalents of $2n$ and 2^{-n} for values of n up to 16.

In an A/D or D/A converter, the first bit is called the most significant bit or MSB and has the weight of $\frac{1}{2}$ full scale of the converter; the second bit has a weight of $\frac{1}{4}$ full scale or FS and so on down to the last bit which is called the least significant bit or LSB and has a weight of $\frac{1}{2}n$FS. The resolution of the converter is determined by the number of bits, and the size of the LSB is FS/2n. It should be noted that the digital code used does not, in general, correspond to its decimal equivalent in analog voltage. The coding used is the set of coefficients of 2^{-n} representing a fractional part of full scale. The MSB is always positioned on the left and the LSB on the right of the digital code. The *binary code* 10110 thus represents $(1 \times \frac{1}{2}) + (0 \times \frac{1}{4}) + (1 \times \frac{1}{8}) + (1 \times \frac{1}{16}) + (0 \times \frac{1}{32})$ or $\frac{11}{16}$ of full scale of the converter. The full scale analog value for a converter can be any convenient voltage, but voltages such as 0 to +5, 0 to +10, ±2.5, ±5 and ±10 are most commonly used. A 12-bit converter, for example, has a resolution of 1 part in 4096. If the full scale analog voltage is 10 volts, then the LSB size is 10 V/4096 or 2.44 *millivolts*.

TABLE 4.1   Decimal Equivalents of 2^n and 2^{-n}

n	2^n	2^{-n}	dB
0	1	1.0	0.0
1	2	0.5	−6.0
2	4	0.25	−12.0
3	8	0.125	−18.1
4	16	0.0625	−24.1
5	32	0.03125	−30.1
6	64	0.015625	−36.1
7	128	0.0078125	−42.1
8	256	0.00390625	−48.2
9	512	0.001953125	−54.2
10	1,024	0.0009765625	−60.2
11	2,048	0.00048828125	−66.2
12	4,096	0.000244140625	−72.2
13	8,192	0.0001220703125	−78.3
14	16,384	0.00006103515625	−84.3
15	32,768	0.000030517578125	−90.3
16	65,536	0.0000152587890625	−96.3

The resolution of a converter can be conveniently related to dynamic range in dB since a factor of 2 corresponds to 6.02 dB. Therefore, the number of bits × 6.02 gives the dynamic range in dB. A 12-bit converter, therefore, has a dynamic range of 72.2 dB.

Converters have both unipolar and bipolar analog values and use a number of different binary related codes. Table 4.2 shows binary coding for a unipolar 8-bit converter with a 10-volt full scale. Notice that all 1's in the digital code do not correspond to full scale but to $(1-2^{-n})$FS.

In some converters it is convenient to use reverse-sense binary coding, or complementary binary, where the most negative analog value corresponds to full-scale digital value. This code is just the binary code with all 1's made 0 and vice versa. For bipolar analog values, the most common codes are off-

TABLE 4.2   Binary Coding for 8-Bit Unipolar Converters

Scale	+10VFS	Straight binary MSB LSB	Complementary binary MSB LSB
+FS − 1LSB	+9.96	1 1 1 1 1 1 1 1	0 0 0 0 0 0 0 0
$+\frac{3}{4}$FS	+7.50	1 1 0 0 0 0 0 0	0 0 1 1 1 1 1 1
$+\frac{1}{2}$FS	+5.00	1 0 0 0 0 0 0 0	0 1 1 1 1 1 1 1
$+\frac{1}{4}$FS	+2.50	0 1 0 0 0 0 0 0	1 0 1 1 1 1 1 1
$+\frac{1}{8}$FS	+1.25	0 0 1 0 0 0 0 0	1 1 0 1 1 1 1 1
+1LSB	+0.04	0 0 0 0 0 0 0 1	1 1 1 1 1 1 1 0
0	0.00	0 0 0 0 0 0 0 0	1 1 1 1 1 1 1 1

TABLE 4.3   Binary Coding for 8-Bit Unipolar Converters

Scale	±5VFS	Offset binary MSB LSB		2's complement MSB LSB	
$+$FS $-$ 1 LSB	+4.96	1 1 1 1	1 1 1 1	0 1 1 1	1 1 1 1
$+\frac{3}{4}$FS	+3.75	1 1 1 0	0 0 0 0	0 1 1 0	0 0 0 0
$+\frac{1}{2}$FS	+2.50	1 1 0 0	0 0 0 0	0 1 0 0	0 0 0 0
0	0.00	1 0 0 0	0 0 0 0	0 0 0 0	0 0 0 0
$-\frac{1}{2}$FS	−2.50	0 1 0 0	0 0 0 0	1 1 0 0	0 0 0 0
$-\frac{3}{4}$FS	−3.75	0 0 1 1	0 0 0 0	1 0 1 0	0 0 0 0
$-$FS $+$ 1 LSB	−4.96	0 0 0 0	0 0 0 1	1 0 0 0	0 0 0 1
$-$FS	−5.00	0 0 0 0	0 0 0 0	1 0 0 0	0 0 0 0

TABLE 4.4   BCD Coding for 2-Digit Unipolar Converters

Scale	+10VFS	BCD MSB LSB	
$+$FS $-$ 1 LSB	+9.9	1 0 0 1	1 0 0 1
$+\frac{3}{4}$FS	+7.5	0 1 1 1	0 1 0 1
$+\frac{1}{2}$FS	+5.0	0 1 0 1	0 0 0 0
$+\frac{1}{4}$FS	+2.5	0 0 1 0	0 1 0 1
$+$1 LSB	+0.1	0 0 0 0	0 0 0 1
0	0.0	0 0 0 0	0 0 0 0

set binary and 2's complement which are shown for an 8-bit bipolar converter in Table 4.3.

Offset binary is simply a shifted binary code where $\frac{1}{2}$FS binary corresponds to *analog zero*. 2's complement coding is the same as offset binary except that the MSB is complemented, resulting in a digital code of all 0's corresponding to analog zero. *Binary coded decimal* or BCD coding is also commonly used in converters and is illustrated by the 8-bit coding shown in Table 4.4. In BCDs, 4 binary bits are used to code each decimal digit. This code can also be used for bipolar analog values if a separate sign bit is used. Other codes such as the Gray code, sign-magnitude binary, and 1's complement are sometimes used but are not as common as the codes just described.

D/A CONVERTERS

In addition to being used as the basis for a large fraction of all A/D converters that are manufactured, D/A converters have a large number of important uses in their own right. Among these uses are *computer*-driven CRT displays, digital-controlled power supplies for automatic test equipment,

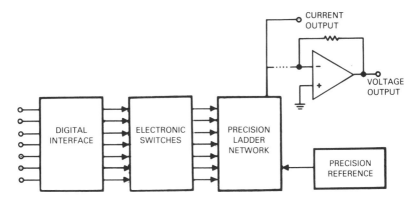

Figure 4.13   Diagram of D/A converter.

digital generation of analog waveforms, and digital control of automatic process control systems.

Although there exist a large array of techniques for accomplishing digital-to-analog conversion, the methods discussed here will be limited to the most commonly used parallel conversion methods. The basic configuration of a D/A converter, or DAC, is shown in Figure 4.13.

A digital interfacing circuit converts the logic inputs to the control levels of a set of switches. These operate in conjunction with a precision *resistor ladder network* to give binary weighted currents or voltages; the ladder network is referenced to a stable precision voltage source. The output of the ladder network is the sum of all the binary weights in the form of a voltage or current. In Figure 4.13, a current output ladder is shown. The two types of D/A converters are current output DACs and voltage output DACs. For current types, the current output of the ladder is brought out as the output of the converter; for voltage output types the current goes to an operational amplifier current-to-voltage converter circuit. In most high-speed applications, a current output DAC is used since there is always some loss of speed due to current-to-voltage conversion.

A frequently used method of achieving a binary weighted set of currents is the circuit shown in Figure 4.14. A series of transistor current sources has its *collector* currents set by *emitter* resistors with values of R, 2R, 4R, 8R, etc. A stable reference voltage, compensated for base-to-emitter voltage variation with temperature, is used to bias the bases of all transistors and thus to set up constant emitter currents. The current source transistors are switched on or off by logic inputs connected through diodes to the emitters. Depending on the logic level at each diode input, the current will flow through either the diode or the transistor. The weighted currents are summed at the collectors of all transistors and become either the current output of the converter or the current input to an operational amplifier as shown in Figure 4.14.

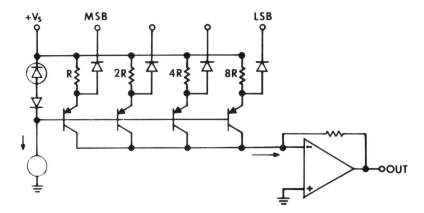

Figure 4.14 Weighted current sources for D/A converter.

Alternatively, to preserve the high speed of the current output, a resistor can be connected to this output current to convert the current directly to a voltage. This is an excellent way of maintaining the speed of the converter but it is restricted to relatively small full-scale voltage swing of the transistor collectors.

The weighted current source method has the advantage of simplicity and high speed. Current output DCA's of this type can also be used in configurations where one or more units to sum their output currents together directly. The disadvantage of this method is the wide range of resistance values required for a high-resolution converter and the resultant effect on both *temperature tracking* and speed. Nevertheless, high-*resolution* converters with high speed can be made by using several groups with 4 or 5 current sources each, and dividing the current output of each group. This is illustrated in Figure 4.15 which shows 3 groups of current sources with resistive current dividers following groups 2 and 3. If each group has 4 binary current sources, then the dividers would have to reduce the current outputs of groups 2 and 3 to 1/16 of their original values.

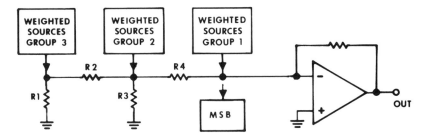

Figure 4.15 Groups of identical binary weighted current sources to achieve high resolution.

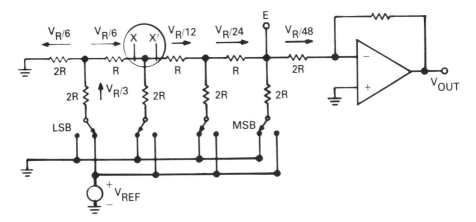

Figure 4.16 R-2R network D/A converter.

Figure 4.15 also shows the method of achieving a bipolar output for a D/A converter. A current source with a current output equal to the MSB is connected to the output of all the other weighted sources. This offsets the output of the converter by one half the full-scale value. This gives offset binary coding as previously discussed.

A second popular method for D/A conversion is the R-2R ladder technique. As shown in Figure 4.16, this consists of a network or series values of R and shunt values of 2R. The bottoms of the shunt resistors are switched between a voltage reference source and common. The operation of the ladder network is based on the binary division of current as it flows down the ladder. This can be seen by examination of the points X and X' in the network. The following conclusions are valid: from point X looking to the right, a resistance R is seen, and looking to the left, a resistance of 2R is seen; from point X' looking to the right, a resistance of 2R is seen, and looking to the left, a resistance of R is seen. These properties hold for any of the junctions along the ladder. If a 2R resistor is switched to the voltage reference source, the source sees a resistance of 2R plus 2R in parallel with 2R or 3R total, and a current of $V_{ref}/3R$ flows into the junction. At the junction this current divides equally with half flowing to the right and half to the left. The right-hand current flows to the next junction where it is again divided in half, and so on to the right end of the ladder where it becomes part of the total output current. The total output current is the sum of all the currents from the shunt resistors which are weighted binarily with the LSB at the leftmost switch and the MSB at the rightmost switch. The output of the ladder can be a current, shown going into the operational amplifier input, or it can be a voltage appearing at terminal E.

An alternate way of driving the ladder is by means of equal value switched current sources feeding each junction. The bottom of the 2R resistors would then be connected to common. The result is exactly the same as with the voltage source, with each current being divided in half at each junction. The advantages of the 2R method are the following:

1. All resistors are values of either R or 2R resulting in easy matching and temperature tracking.
2. Resistor values can be kept low to insure high speed.
3. The output amplifier always sees a constant resistance value at its input terminal.

Compared to the weighted current source method, the R-2R ladder requires two resistors per bit whereas the former requires one resistor per bit.

A/D CONVERTERS

In analog to digital conversion there is an even greater number of methods commonly used than in D/A conversion. This is so because A/D conversion, except in the very fastest requirements, lends itself to many serial methods and to indirect methods. The discussion here will be limited to four widely used types of A/D converters that are available in low cost modules. These types are:

1. Counter or servo type
2. Dual slope integration type
3. Successive approximation type
4. Parallel type

Counter or Servo Type

This type of A/D converter is one of the simplest and least expensive to implement. The converter is illustrated in simplified form in Figure 4.17. At the start of conversion the clock is gated on, and the digital counter begins to count clock pulses. As it counts, it changes the output of the D/A converter which is compared to the analog input voltage. When the DAC output is equal to the input, the comparator changes state and *inhibits* the clock pulses. At this time conversion is complete and the output digital number is contained in the output *register* of the *counter*.

This converter features simplicity, low cost, and good accuracy but has the disadvantage of slow speed. Conversion time is proportional to input voltage and is longest for full-scale input. In some applications the speed is improved if an up-down counter is used, and the converter counts either way up or down from its previous value rather than resetting to zero. In this way the counting converter can follow slowly varying inputs. This converter is

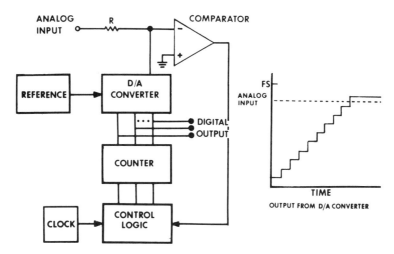

Figure 4.17   Counter or servo A/D converter.

also called a servo type because of the feedback method of controlling the counter.

Dual Slope Integrating Type

Integrating A/D converters operate by the indirect method of converting a voltage to a time period which is then measured by a counter. There are several different types of converters using the *integrating* or ramp principal including single ramp, dual ramp, and triple ramp, which are all variations of the basic principle. The most popular and widely used at present is the dual ramp, or dual slope type, shown in Figure 4.18, which is used in most digital *voltmeters* and digital *panel meters*. The advantages of this method are relatively low cost, simplicity, high accuracy and linearity, and with excellent noise rejection characteristics.

Conversion starts with the unknown input voltage switched to the integrator input. When the output ramp crosses the comparator threshold, the clock is *gated* to the counter which counts up to a predetermined number. At this time the input of the integrator is switched to the reference and the counter reset to zero. The integrator then integrates the reference back down to the comparator threshold, at which time the count is stopped. The input voltage is then the ratio of T2 counts to T1 counts times the reference voltage, and can be read directly from the counter register.

The dual slope method has a number of important features. Conversion accuracy is dependent on the *clock frequency* and integrating capacitor value as long as they are stable within a conversion period, and depends only on the accuracy and stability of the reference. Resolution is basically limited only by the analog resolution of the converter. In addition, this converter gives excellent noise rejection because of the integrating operation; in par-

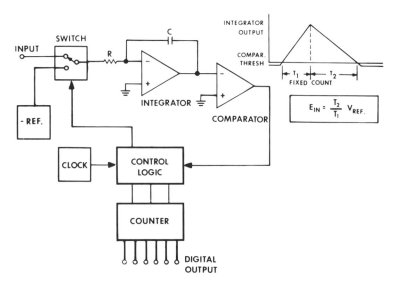

Figure 4.18 Dual slope A/D converter.

ticular, normal mode noise rejection is finite for T1 equal to a multiple of the period of the interfering noise. The main drawback of this method is the relative long conversion time.

Successive Approximation Type

This conversion method is the most widely used in general practice due to its combination of high resolution and high speed. The successive approximation converter operates with a fixed conversion time per bit, independent of the value of the analog input. The method is illustrated in Figure 4.19 and

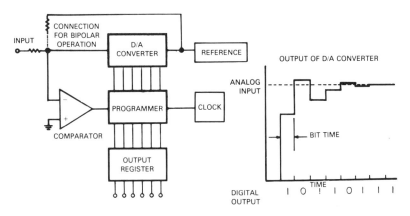

Figure 4.19 Successive approximation A/D converter.

operates by comparing the input voltage with the D/A converter output, one bit at a time.

At the start of the conversion cycle, the D/A converter's MSB output, which is $\frac{1}{2}$ full scale, is compared with the input. If it is smaller than the input, the MSB is left on and the next bit is tried. If the MSB is larger than the input, it is turned off when the next bit is turned on. This process of comparison is continued down to the LSB after which the output register contains the complete output digital number. Both serial and parallel output data can be brought out of this converter and, in addition, the conversion can be synchronized to an external clock on some units. Speeds as high as 100 nanoseconds per bit can be achieved by this method. Successive approximation converters can also be quite accurate, but the accuracy depends on the stability of the reference, the switches, the ladder network, and the comparator. Figure 4.19 also shows how bipolar operation is accomplished using a precision resistor connected to the reference source to the comparator input, thus subtracting a $\frac{1}{2}$ full scale current from the input. Many converters have this resistor built in so that bipolar operation can be achieved by external pin connection.

Parallel Type

While the successive approximation converter is capable of speeds as high as 100 nanoseconds per bit, giving conversion rates of 1 MHz for 10 bits, significantly faster conversion requires the parallel technique. This method is sometimes referred to as the simultaneous, or flash technique, and is capable of 25 MHz conversion rates for 4 bits. As shown in Figure 4.20, the method employs an input quantizer comprised of $2n - 1$ comparators biased 1 LSB apart by a reference voltage. For a given analog input voltage to the comparators, all comparators below the input level turn on while all comparators above it are off. The quantization process is accomplished in the switching time of a single comparator.

The comparator outputs, however, are not in binary code and must, therefore, go through a *decoder*. The parallel method has the advantage of the fastest speed but is limited to a relatively few bits, usually about 4, due to the large numbers of comparators required. To convert a larger number of bits, it is necessary to employ a hybrid technique whereby a parallel conversion stage is followed by a fast D/A converter, the output of which is subtracted from the input voltage. The difference is amplified and then converted using another parallel stage. This results in a speed compromise but higher resolution.

There are a number of important parameters used to characterize the accuracy of A/D and D/A converters. Since they have specific meanings as applied to converters, these parameters are defined here.

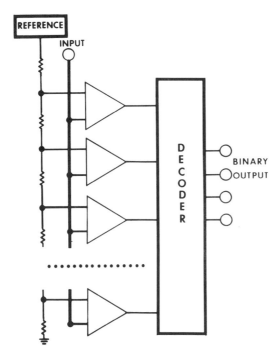

Figure 4.20 Parallel type A/D converter.

Resolution. The smallest analog change that can be distinguished by an A/D converter or produced by a D/A converter. Resolution is the analog value of the LSB, which is $FS/2n$ for an n-bit binary converter and $FS/10d$ for a d-bit BCD converter. Resolution is often specified in percentage of full scale, as in the 10-bit converter which has a resolution of 0.1%. In many cases, the useful resolution of a converter may be less than the specified resolution.

Linearity. The maximum deviation from a straight line drawn between end points of the converter transfer function. Linearity may be expressed as a percentage of full scale or as a fraction of LSB size. The linearity of a good converter is $\pm\frac{1}{2}$ LSB, as shown in Figure 4.21.

Differential linearity. The maximum deviation of an actual bit size from its theoretical value for any bit over the full range of the converter. A differential linearity of $\pm\frac{1}{2}$ LSB means that the size of each bit over the range of the converter is 1 LSB $\pm\frac{1}{2}$ LSB as shown in Figure 4.22.

Monotonicity. Having a continuously increasing output for a continuously increasing input over the full range of the converter. Monotonicity requires that the differential linearity be less than 1 LSB (see Figure 4.22).

Missing code. This occurs in an A/D converter when the output code skips a digit. It happens when the differential linearity is greater than 1 LSB for some bit.

Quantizing error. The basic uncertainty associated with digitizing an analog signal due to the finite resolution of an A/D converter. An ideal converter has a maximum quantizing error of $\frac{1}{2}$ LSB (refer to Figure 4.2).

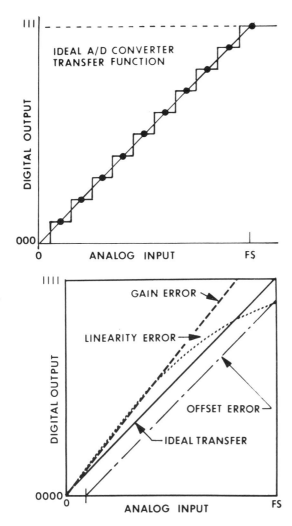

Figure 4.21   Gain, offset, and linearity errors.

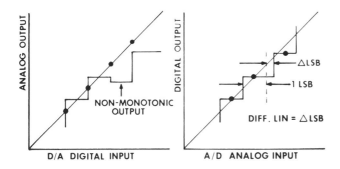

Figure 4.22   Monotonicity and differential errors.

Relative accuracy. The input to output error as a fraction of full scale, with gain and off-set errors adjusted to zero. Relative accuracy is a function of linearity.

Absolute accuracy. The full-scale analog error referenced to the NBS standard volt.

Offset error. The error by which the transfer function fails to pass through the origin, referred to the analog axis. This is adjustable to zero in available converters (see Figure 4.21).

Gain error or scale factor error. The difference in slope between the actual transfer function and the ideal function in percent. This error is adjustable to zero in available converters (see Figure 4.21).

ANALOG MULTIPLEXERS

Analog multiplexers are used for *time sharing* of analog to digital converters between a number of different analog information channels. An analog multiplexer consists of a group of analog switches arranged with inputs connected to the individual analog channels and outputs connected in common, as shown in Figure 4.23. The switches can be addressed by a digital input code. MOSFET switches are generally used and may be connected directly to an output load if it is a high enough impedance, or to an output buffer amplifier which provides a very high impedance to the switches. Using a fast bipolar transistor-following amplifier as a buffer, an input impedance of 1000 megohms is achieved, resulting in a negligible transfer error due to the switch resistance of typically 2k ohms.

Figure 4.23 shows the equivalent circuit of one of the MOSFET switches which is characterized by series and shunt resistance, shunt capacitance, and a leakage current. Differential analog multiplexing can also be accomplished by using two MOSFET switches for each input channel.

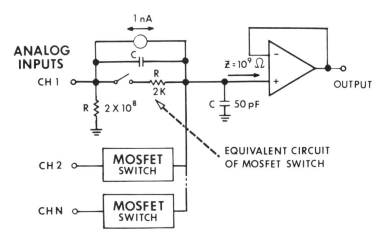

Figure 4.23 Analog multiplexer circuit.

There are several important parameters used to characterize analog multiplexers that are defined as follows:

Transfer accuracy. The input-to-output error as a percentage of the input. Transfer accuracy depends on the source impedance, switch resistance, load impedance (if the multiplexer is not buffered), and the signal frequency.

Settling time. The same definition as discussed earlier. Here it includes the switching time of the switches.

Throughput rate. The highest rate at which the multiplexer can switch from channel to channel at its specified accuracy. This rate is determined by the settling time.

Crosstalk. The amount of signal coupled to the output as a percentage of input signal applied to all OFF channels together. The inverse of this percentage is expressed as attenuation in dB.

Input leakage current. The highest current that flows into or out of an OFF channel input terminal due to switch leakage.

SAMPLE AND HOLD CIRCUITS

Some of the properties of sample and hold circuits were discussed in the section on sampling theory. Here the circuit configurations and operating parameters will be discussed. Sample and hold circuits are used in conjunction with both A/D and D/A converters. With A/D converters, they are used to shorten the aperture time for the converter by rapidly sampling the input signal and then holding its value until the conversion is completed. With D/A converters, these circuits are used in display applications to remove "glitches" which appear at the output of all DAC's as they change from one analog level to another.

As described before, in its basic form, a sample and hold consists of a switch and a capacitor. When the switch is closed the unit is in the sampling or tracking mode and will follow a charging input signal. When the switch is opened the unit is in the hold mode and will retain a voltage on the capacitor for some period of time depending on capacitor and switch leakage.

Practical sample and hold circuits also use input and output buffer amplifiers and sophisticated switching techniques. The output buffer amplifier must be a low input current FET amplifier in order to have as small an effect as possible on the leakage of the capacitor. Similarly, the electronic switch used must be a low leakage type such as an FET switch. Figure 4.24 illustrates two commonly used sample and hold configurations. Circuit (a) is

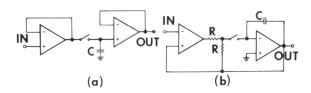

(a)   **(b)**

Figure 4.24   Sample and hold circuits.

used for fast sample and hold and is an open loop configuration using fast voltage follower amplifiers. For very fast circuits, a diode bridge–type sampling switch is used.

Circuit (b) is a closed loop configuration with an operational integrator in the feedback path of the input buffer amplifier. This circuit results in extremely good accuracy and linearity.

Sample and holds are characterized by a number of important parameters, each with a specific meaning for these circuits and defined as follows:

Acquisition time. The time from when the sample command is given to the point when the output enters and remains within a specified error band around the input value. At the end of the acquisition time the output is tracking the input. Note a similarity to the definition of settling time.

Aperture time. The time lapse between the hold time and the point at which the sampling switch is completely open. Aperture time is also referred to as turn-off time.

Aperture uncertainty time. The variation in aperture time for sample and hold. The difference between maximum and minimum aperture time.

Decay rate. The maximum change in output voltage with time in the hold mode.

Feedthrough. The amount of input signal appearing at the output when the unit is in the hold mode. Feedthrough varies with signal frequency and may be expressed as an attenuation in dB.

A/D AND D/A CONVERTER ADJUSTMENTS

A timing diagram for a typical successive approximation A/D converter is shown in Figure 4.25. The start of conversion is initiated by a start convert pulse of 30 nanoseconds minimum duration. The EOC (end of conversion) or status output then goes high indicating that conversion is in process. The MSB output of the D/A converter is turned on to be compared with the input voltage, registering a logic "1" at the Bit 1 parallel data output. If the MSB is smaller than the input voltage, then it stays on and the logic "1" remains on Bit 1 output. If the MSB is larger than the input, then it turns off after the first clock pulse and Bit 1 output goes to a logic "0". After the first clock pulse, the serial output for Bit 1 appears on the serial output pin. Then Bit 2 is compared with the input remainder, after the second clock pulse Bit 3 is compared, and so on down to Bit 10 in this case. Thus each succeeding bit is compared to the remaining input during one clock interval, and the true output for that bit appears at the next clock interval. At the end of the tenth clock pulse, all parallel bit outputs are true and the serial output appears. Also, the EOC output returns to a logic "0" indicating that conversion is complete.

The EOC or status output may be used to control the sample and hold preceding the A/D converter, since when it is low, the sample and hold is put

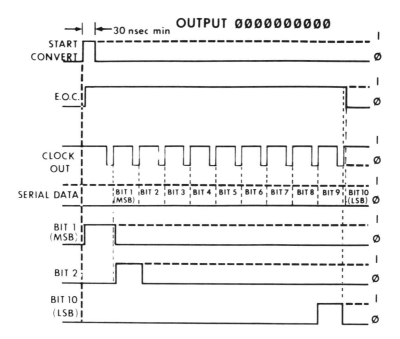

Figure 4.25   Timing diagram for 10 bit A/D converter.

in the tracking mode, and when it is high, during conversion, the sample and hold is put in the hold mode.

Most Datel A/D and D/A converters are capable of both unipolar and bipolar operation with several analog ranges selected by external pin connection. The commonly used ranges are 0 to +5 volts, 0 to 10 volts, ±2.5 V, ±5 V, and ±10 V. The digital coding used is binary or BCD for unipolar operation and offset binary or 2's complement for bipolar operation. All Datel A/D and D/A converters have provisions for user adjustment of full scale and offset in order to maintain optimum accuracy in a given operation. Figure 4.26 shows the circuit connection necessary to perform the calibration adjustments on an A/D converter.

Calibration is accomplished for a unipolar A/D converter as follows:

1. Connect a precision pulse generator to the "Start Convert" input terminal. The generator should be set to give a pulse width and amplitude as specified in the converter data sheet. The repetition rate should be set for the conversion time of the converter.

2. Connect a precision voltage reference source to the analog input terminal.

3. Adjust the output of the voltage reference source to zero plus $\frac{1}{2}$ LSB. This voltage is found by multiplying the converter full-scale voltage range by $\frac{1}{2}$ the value of 2^{-n} for the n-bit converter. Adjust the zero

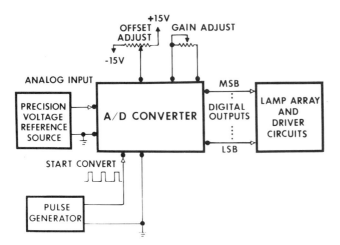

Figure 4.26 Connections for calibrating of A/D converter.

offset trimming potentiometer until the LSB output flickers between logic "0" and logic "1".

4. Adjust the output from the voltage reference source to full scale minus $1\frac{1}{2}$ LSB. Adjust the gain trimming potentiometer until the LSB output flickers between logic "0" and logic "1".

In the case of a bipolar A/D converter, the calibration procedure is identical to the unipolar calibration procedures except that the first part of step 3 is modified as follows:

3. Adjust the output from the voltage reference source to *minus full scale plus* $\frac{1}{2}$ *LSB.*

Calibration of D/A converters is similar but even simpler. A zero digital input is applied to the converter and the offset trimming potentiometer is adjusted to give zero analog output. Full-scale digital input is applied and the gain trimming potentiometer adjusted to give an analog full-scale output minus 1 LSB.

CHAPTER REVIEW QUESTIONS

1. What type of control function requires analog control signals and feedback?
2. How can analog signals be processed by a digital computer?
3. What is a binary code?
4. How does BCD differ between the basic binary codes?
5. What is a low-pass filter?
6. How does a low-pass filter aid in reducing frequency-fold errors?

7. What is the purpose of a buffer circuit?

8. What is the maximum gain of a unity amplifier?

9. What data information is contained in the MSB?

10. What does the term *impedance* mean in relation to an analog signal or frequency?

11. Define the difference between an A/D converter and a D/A converter.

12. Which type of operational amplifier produces the highest gain: a closed-loop or an open-loop amplifier circuit?

13. What type of A/D converter provides the combination of high-resolution and high-speed conversions?

14. How much time does the A/D converter in Question 13 require to process 1 bit?

15. What technique offers 25-MHz conversion speeds for 4-bit data?

GLOSSARY

Acquisition time. Time period necessary to acquire or receive a given amount of data. Stated in milliseconds, microseconds, or nanoseconds.

A/D converter. Analog-to-digital converter.

Alias. An error signal created by an insufficient sampling rate of an analog waveform.

Amplifier. An electronic circuit used to increase power, voltage, or current from a low-level input.

Amplitude. The maximum displacement of an analog signal above or below analog zero. The maximum deviation of an alternating current or voltage during one cycle of the alternating frequency.

Analog signal. A signal consisting of a continuously varying or randomly varying amplitude above or below analog zero.

Analog-to-digital converter. A device by which an analog signal input can be converted to a binary digital representation of the analog signal data.

Analog zero. A point between the positive amplitude and the negative amplitude of an analog signal where the analog signal is neither positive nor negative. Analog zero occurs at 0, 180, and 360° during one 360° cycle of a sine wave representing analog data.

Aperture time. The time to make one conversion of an analog input signal.

Attenuator. A device by which an analog signal amplitude, voltage, or current can be reduced to a desired level.

Bandwidth. A minimum range of frequencies necessary to transmit analog data in a specific period of time.

BCD. Binary-coded decimal.

Binary code. A numerical system to the base 2 representing a numerical value from a system to the base 10.

Binary coded. A four-digit binary code representing decimal numbers from 0 to 9. A separate four-digit code is used to represent units, tens, hundreds, thousands, and so on, for any decimal number.

Bipolar. Having two poles, such as the N and S poles of a magnet; the polarity of P and N semiconductor compositions of a transistor, diode, or integrated circuit.

Boost. To boost or increase the amplitude, voltage, or current of an electrical signal by amplification of the desired factor.

Buffer. A nonloading device or state that isolates one circuit from another to prevent degrading one circuit by the coupling effects of a following circuit.

Capacitor. An electronic component capable of storing a signal current or voltage for a time period limited by the leakage current of the capacitor and the capacitor input device.

Clock. An electronic circuit designed to generate a given frequency of timing pulses; a device capable of displaying real time over a period of time denoting a 24-hour period or day.

Clock frequency. The number of clock pulses generated in a period of 1 s; the frequency of the clock's oscillator frequency control crystal.

CMRR. Common-mode rejection ratio.

Collector. A terminal of a transistor which is normally connected to the source voltage through a load resistor; the destination of the majority of the electrons passing through an NPN transistor emitter circuit.

Computer. An electronic device with the capability of accepting input data to be processed by arithmetic or logic functions, stored, retrieved, and returned to the user as a display on the screen of a CRT.

CRT. Cathode ray tube.

D/A. Digital to analog.

DAC. Digital-to-analog converter.

Data transmission link. A connecting path of wire conductors or cable by which a computer may transmit data bidirectionally between internal sections of the computer or to external peripherals.

dB. Decibel.

Decoder. A device by which data in one form of a numerical code are translated into another numerical form for processing by a computer.

Dielectric. An insulating, nonconductive material that maintains electrical separation between the plates or foil sections of a capacitor (e.g., air, mica, glass, wax, oil, plastic, ceramics, or chemical foam).

Digital code. A number system to the base 2 used to express digital logic states by voltage levels, where a low voltage, 0 through 0.2 V, denotes a logic 0 and a high voltage, 3.4 to 5 V, denotes a logic 1.

Diode. A solid-state semiconductor device which has the property of passing current in one direction only—from cathode to anode.

Distortion. A change in a linear waveform that results in a nonlinear waveform with a loss of a portion of information contained in the original waveform.

Emitter. The element or section of a transistor from which electrons leave and move toward the collector.

Filter. An electronic circuit or device to attenuate or emphasize specific frequencies; a low-pass filter produces low attenuation to a range of frequencies below a cutoff point and high attenuation to those frequencies above the cutoff point.

Frequency. The number of cycles of an alternating-current or analog waveform occurring in 1 s; the number of digital pulses generated by a clock in 1 s; a measure of bandwidth or rate of data transfer.

Frequency folding. An effect produced when the sampling rate is less than the analog frequency and causes some of the higher frequencies to fold back into the signal, producing distortion and loss of analog data information.

FS. Full scale.

Impedance. The vectorial sum of all resistance and reactance in an analog or alternating-current circuit.

Inductor. An electronic component with a current path in the form of loops or coils which has the property of producing a magnetic field, surrounding the device, when current passes through its coils of wire, thus opposing any change in current intensity by inducing current back into the current path when a reduction of current results in a partial collapse of the surrounding magnetic field.

Inhibit. To stop, halt, interrupt, or prevent activation of a device or transmission of data until certain conditions or time are met.

Instrument amplifier. Typically, an operational amplifier operated in a feedback or open-loop mode for high amplification of an analog instrument output.

Irreducible. Cannot be further reduced; cannot be eliminated.

Link. To join together as in a chain; to connect two electronic circuits together via wire or cable; to tie several computer functions together in series or parallel configurations.

Low-pass active filter. A filter designed to pass all frequencies below a cutoff frequency and reject all frequencies above the frequency of cutoff.

MHz. Megahertz.

Microsecond. One millionth of a second; 0.000001 s.

Millisecond. One thousandth of a second; 0.001 s.

Modulation. The process by which two ranges of frequencies may be combined into a single range of frequencies which on demodulation can restore both ranges of frequencies to their original states (e.g., radio frequencies modulated by audio frequencies, television frequencies modulated with video and audio frequencies).

Modulation envelope. Amplitude variations produce in the amplitude of radio carrier frequency when modulated with a lower-frequency analog signal.

MOSFET. Metal-oxide silicon field-effect transistor.

Nanosecond. One billionth of a second; 0.000000001 s.

Nonlinearity. Deviation from a straight plane; logarithmic output variations from a linear input; distortion resulting from nonlinear amplification.

ns. Nanosecond.

Offset binary. An 8-bit binary number with one digit shift to right or left of its original position.

Op-amp. Operational amplifier.

Panelmeter. A dial or digital display device for indicating voltage, current, power, time, or frequency and installed on an equipment rack panel rather than separately case mounted.

Parameter. One of the independent variables in a set of equations. A determining factor or characteristic. A constant or variable term of a function that determines the form, but not the nature, of the function.

Pulse. A variation in a voltage level that lasts for a brief period of time. A train of equally spaced pulses with a duration limited by the frequency of the generating circuit or device (e.g., timing clock pulses, momentary pulse, or single-pulse trigger or reset pulse).

Quantization register. To restrict a variable to discrete values, each of which is an integral multiple of the same quantity. A section of a microcomputer that momentarily receives input data to be manipulated, transfers the data to be manipulated, or holds program instruction addresses.

Resistor. An electronic component having the property of limiting the flow of current in an electronic circuit. Also referred to as a bias resistor, load resistor, pull-up resistor, and in network configuration—an integrating resistor.

Resistor ladder network. A series of resistors connected in a ladder form, with one side of the network at ground potential.

Resolution. The smallest incremental step in separating a measurement into its constituent parts. In a digital system, resolution is one count in its least significant digit.

RLC. Resistor-inductor-capacitor network.

Sample and hold. To take a measure of voltage at different intervals and hold until the next sample is taken.

Sawtooth. A waveform where the amplitude rises at one angle and declines at another angle; a triangular-shaped waveform.

Sequential. One after another, as the number of pages in a book.

Settling time. The time lapse between the application of a full-scale step input to an amplifier until the output has entered and remains within a specified error band.

Signal current. The increase or decrease of electrons moved by an increase or decrease of signal amplitude voltage.

Slew. The rise-time slope of an amplifier receiving a maximum square-wave input signal.

Switch. A mechanical device for making or breaking circuit current paths in an electrical or electronic circuits.

Temperature tracking. Characteristic operational amplifier variations resulting from ambient temperature changes. This effect can be compensated with temperature-compensating components.

Train. A series of pulses transmitted one after another. A series of data bits sent single file or serially.

Transducer. A device that receives energy in one form and retransmits it in another form (e.g., a microphone receives sound waves and changes them to electrical analog waves).

Transistor. A solid-state device with the ability to control large current flows from small-signal inputs. Its name comes from the function it performs, that of a *transfer resistor.*

Unipolar. Having a single pole or a single charge carrier.

Unity. A single entity, a gain of 1.

Voltage. An electromotive force; difference of potential; a positive attraction for negative current.

Voltmeter. A measuring device for measuring the effects or magnitude of various voltage levels.

V_{ref}. Reference voltage.

5
CMOS Technology

*The technology of microelectronics is advancing so rapidly
that by the time a new device reaches the production state it is obsolete.*

The integrated-circuit (IC) processing technology, in its relative short history, has undergone very rapid changes. It is interesting to look back in time at the major turning points in the evolution of the integrated circuits.

The first integrated circuits were manufactured by the bipolar process. As this early technology improved, the ICs evolved from small-scale integration (SSI) with only a few gates in one package, to chips that were larger and more complex to medium-scale integration (MIS). As the need developed for even greater circuit density, large-scale integration (LSI) was implemented by integrating entire subsections of a digital system on a single chip.

In order to meet yield and manufacturing objectives, it soon became apparent that a process simpler than the bipolar process had to be developed. Thus began the first of a long series of metal-oxide semiconductor (MOS) processes, out of which the P-channel or PMOS process was developed.

It is not often realized by newcomers to electronics that the PMOS processes were used to produce nearly all of the early LSI chips. It was used to produce the first semiconductor memory chips, the first calculator chips, and the first microprocessor chips. The PMOS processes produced higher circuit density, higher yields, and lower-cost ICs because they required fewer process steps than the bipolar process.

A cross section of a PMOS transistor is shown in Figure 5.1. Note that only a single P^+ diffusion was used and this formed both source and drain

*Material in this chapter has been adapted from various *National Semiconductor* publications.

116

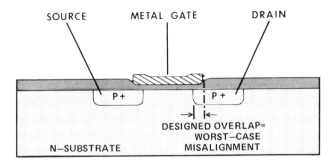

SOURCE   METAL GATE   DRAIN

DESIGNED OVERLAP=
WORST—CASE
MISALIGNMENT

N—SUBSTRATE

Figure 5.1 PMOS process. (Courtesy of National Semiconductor.)

regions of the transistor. Only five masks were required. There was no epi-layer, no N$^+$ buried layer, and no isolation diffusion. The simplicity of the process resulted in significant higher yields for these early LSI chips.

The metal gate was purposely extended over the source and drain diffusions to ensure that a channel could be created even with mask misalignment. This caused an increase in the drain-to-gate input capacitance and slowed down circuit response time because of the *Miller effect*, which causes input capacitance to increase in apparent value by an amount equal to the voltage gain of the transistor.

The first use for a polysilicon layer (now simply called poly) deposited on the oxide surface of a silicon wafer was to reduce the gate overlap capacitance that existed with the metal-gate PMOS process. Figure 5.2 illustrates the polysilicon-gate technique, which called for putting the poly layer in place first and then, with the help of a silicon-nitride layer on top of the poly, use this poly gate as a diffusion mask to define the channel length of the transistor. By using this self-aligning technique, the input capacitance was reduced by a factor of 5:1, which resulted in smaller chips and higher-speed circuits. The poly layer was heavily doped and served as an additional circuit interconnect layer.

It has always been known that the N-channel MOS transistor is capable of higher-frequency operation than a P-channel transistor; electrons (the current carrier for the N-channel transistor) move faster than holes (the current carrier for P-channel transistors). The problem was sodium contamina-

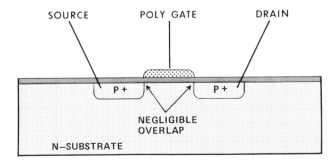

SOURCE   POLY GATE   DRAIN

NEGLIGIBLE
OVERLAP

N—SUBSTRATE

Figure 5.2 Silicon gate process. (Courtesy of National Semiconductor.)

tion during the processing, which caused an increase in the threshold voltage of the P-channel transistor. The high power supply voltages for P-channel transistors were capable of accommodating some degree of the threshold voltage shift, but the threshold voltage of the N-channel transistor would drop to zero and the transistor could not be turned off. Thus it was necessary to develop better knowledge of MOS surfaces and process control before the more difficult NMOS products could be fabricated.

After the NMOS processing problem of sodium contamination was solved, there was a large improvement in circuit performance. Not only was the NMOS transistor basically capable of higher speeds, but it also provided larger currents for a given geometry than could be obtained with PMOS transistors. These circuit improvements increased the capabilities of the microprocessor and started a sequence of ever-increasing bit capability in the dynamic random access memory (DRAM) products.

Mask-making improvements made it possible to scale devices to smaller and smaller sizes. The industry went from the use of mils to microns [micrometers (μm); 0.000001 m = 1 micron) where 1 mil = 25 microns in a new pseudometric system (the last 0.4 micron was dropped to make the problem of scaling easier). The NMOS process opened the door to very large scale integration (VLSI). However, the large, very dense VLSI chips created power problems, and power dissipation became the limiting factor in VLSI circuits.

From the very beginning of solid-state electronics, the undesirable high-current drain of the IC has been a tough challenge for system designers. While the bipolar circuits achieved good performance, they consumed a great deal of power. Large investments in the NMOS technology had created a price/performance trade-off that was hard to beat. Then came CMOS. This technology not only provided the higher speeds needed for the present systems, but it also enabled the components to operate at lower power. The corresponding reduction in power supply requirements contributed to lower overall system cost, and the resulting cooler IC operating temperatures meant a higher level of reliability.

CMOS CIRCUIT PERFORMANCE

Low power dissipation in CMOS logic circuits is achieved by the use of a process that simultaneously provides both P-channel and N-channel MOS transistors. This complementary MOS (CMOS) technology allows the design of circuits that do not consume direct-current (dc) power except for the minor dc leakage current of the P-N reversed-bias junctions. The only time that power is consumed is during the charging and discharging of stray capacitance and during the short time intervals while input voltage passes between logic levels, causing both P and N transistors to conduct momentarily.

Let's look at an example of how CMOS works in a logic circuit. The

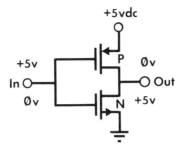

Figure 5.3   A logic inverter illustrating the zero dc power-drain benefit of CMOS circuits. (Courtesy of National Semiconductor.)

CMOS inverter circuit is shown in Figure 5.3 and indicates the lower-power-drain benefits of CMOS logic. Notice that if the input voltage is a logic ∅ (0 V dc), the lower N-channel resistor is OFF and only the upper P-channel transistor is ON. In this state, the output voltage will go to +5 V dc, which is a logic 1 state. If the input changes to a logic 1 state (+5 V dc) the conduction state of the P and N transistors interchange: the P-channel transistor will be OFF and the N-channel transistor will be ON. This pulls the output voltage down to 0 V dc and the output becomes a logic ∅ state. Thus, in both of these steady-state logic conditions, there is no dc power drain.

CMOS technology has not been limited to logic circuit applications; it has become the technology choice for many new linear MOS devices and circuits. For linear IC designers, the CMOS process is the closest to the familiar bipolar process. Many devices are available: P-channel, N-channel, parasitic NPN bipolar transistors, and even zener diodes. The latest concept in linear designs makes use of the excellent analog switch inherent in the CMOS process. Linear designers can make use of both N-channel and P-channel MOSFETs (metal-oxide semiconductor field-effect transistors) in designs that are similar to the NPN and PNP circuits that have long been used for bipolar linear circuits. For a P-well CMOS process, an NPN parasitic bipolar transistor also exists that has a high-current-gain characteristic and can handle large currents. This transistor is restricted to emitter-follower applications because its collector lead is not available; it ties directly to the V_{cc} power supply line. The low output impedance of the emitter follower is often useful as a bias voltage reference within an MOS op-amp design. The base–emitter, forward-voltage drops of these transistors are also useful voltage references, and even bandgap voltage references are possible.

The low-power-drain benefit of CMOS devices makes this technology the only way to achieve certain VLSI functions. The zero-power-drain benefit has also made CMOS the only way to obtain static RAMs that can be "put to sleep." Further advances in the CMOS technology have produced multiple poly layers (P2CMOS) and multiple metal layers (M2CMOS) which offer speeds up to five times faster than the metal-gate CMOS, a 50% reduction in chip area, and power dissipations as little as one-eighth that of the earlier technology.

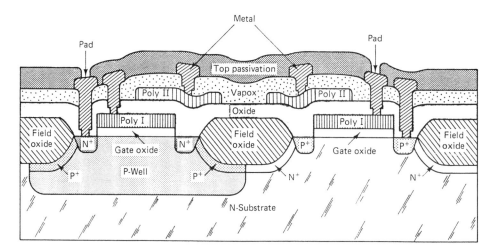

Figure 5.4 Cross section of a logic inverter substrate using the P2-CMOS process. (Courtesy of National Semiconductor.)

THE P2CMOS PROCESS

The P2CMOS process uses two levels of polysilicon and one level of metal for interconnects. A cross section of a CMOS inverter is shown in Figure 5.4. All junctions are formed with ion implantation. High-resolution printing of the images on the wafers and dry etching techniques are used to obtain state-of-the-art feature sizes. Low stray capacitance (achieved by tighter layout rules) allow high-speed performance. Short channel lengths and thin gate oxide layers provide high transconductance in the transistors. All these per-

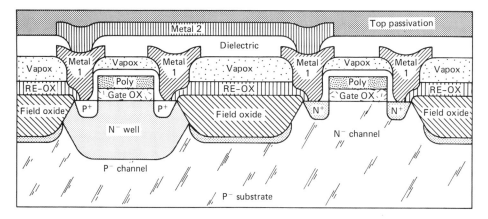

Figure 5.5 Cross section of a logic inverter substrate using the M2-CMOS process. (Courtesy of National Semiconductor.)

formance benefits exist while retaining all the advantages of CMOS. The polysilicate layer, which results when silicon is deposited from the vapor phase onto an oxide surface, is subsequently heavily doped to reduce resistance of the interconnect lines. Sheet resistances of 30 Ω/cm^2 are achieved, which can be further reduced to as low as 2 Ω/cm^2 by adding special metals to the poly layer to form polysilicides.

Interconnect resistance still causes excessive propagation delays in high-speed logic circuits. The two-layer metal process, M2CMOS (illustrated in Figure 5.5), makes it possible to overcome this limitation. With the M2CMOS process, sheet resistance of the metal interconnects can be reduced to only 0.03 Ω/cm^2, 1000 times less than that achieved with doped poly. Notice that Figure 5.5 shows the second metal connecting to the first metal through vias or holes that are etched in the insulating oxide separating the two metals.

SOLVING THE CMOS LATCH-UP PROBLEM

Latch-up is a regenerative phenomenon that occurs where the voltage at an input or output pin is raised above the power supply voltage by a small amount. When this happens, the IC's current becomes higher and higher until, in effect, it virtually shorts V_{cc} to ground, which can destroy the device.

The CMOS process that simultaneously provides both N-channel and P-channel transistors also provides a few parasitic bipolar transistors, both PNPs and NPNs. As long as the voltages applied to the CMOS IC remain within the bounds of the power supply voltages (0 to +5 V dc), these bipolar transistors will remain dormant. Applying negative voltages or voltages larger than the power supply voltage can "wake up" these parasitic bipolar devices and create some major problems.

The regenerative circuit in Figure 5.6 forms the silicon-controlled rectifier (SCR), the "latching element" of the CMOS latch-up problem. When both NPN and PNP bipolar transistors exist on the same IC chip, as in the case of CMOS and all bipolar ICs, the possibility of inadvertently ending up with a particular parasitic interconnection of a NPN with a PNP transistor (called the "hook" connection) has to be considered in the IC chip design

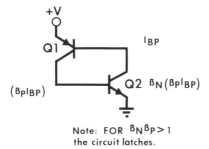

Note: FOR $\beta_N \beta_P > 1$
the circuit latches.

Figure 5.6 "Hook" or SCR connection of complementary bipolar transistors. (Courtesy of National Semiconductor.)

stage. Traditionally, systems designers have solved the latch-up problem by adding external protective circuitry.

National Semiconductors, Inc. has developed proprietary processing and layout enhancements to make their HC logic chips latch-up-proof, the first CMOS logic products to make this claim.

HOW SPEED HAS IMPROVED

National Semiconductors, Inc. uses self-aligned silicon-gate CMOS with oxide isolation to obtain speed and density advantages. To understand how this works, look at the cross section of a logic inverter in metal-gate CMOS shown in Figure 5.7. Notice that all the key parasitics have been drawn on the diagram. The extra P$^+$ diffusions around the N-channel device and the extra N$^+$ diffusions around the P-channel device are guard rings (channel stoppers) that are necessary to prevent parasitic MOS transistors between devices. The length indicated for laying out a logic inverter (one N-channel and one P-channel device) is about 125 microns for a typical metal-gate process.

Now look at the silicon-gate CMOS inverter shown in Figure 5.8; the 125 micron distance previously noted for a simple logic inverter has been reduced to 64 microns. With oxide isolation performing the channel stop function, the need for extra channel stop diffusions is eliminated; the P$^+$ and N$^+$ diffusions can now butt directly against the oxide (no space is required); due to smaller junction depths, the geometries and spacings overall can be made smaller.

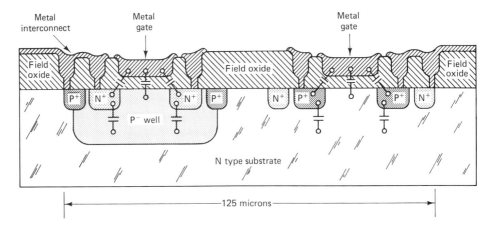

Figure 5.7 Cross section of a logic inverter substrate using the metal-gate CMOS process. (Courtesy of National Semiconductor.)

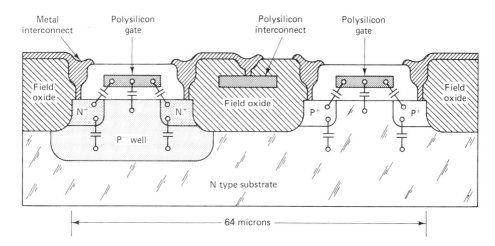

Figure 5.8 Cross section of a logic inverter substrate in P2CMOS. (Courtesy of National Semiconductor.)

A 2:1 reduction in the parasitic capacitance can be obtained with the P2CMOS process. Also, since the gate is self-aligned to the source and drain (i.e., the gate actually defines the separation between the source and drain regions), the gate overlap capacitance is significantly reduced. This decreased capacitance, combined with the fact that the gains of the silicon-gate transistors are increased by a 4:1 ratio (the silicon-gate devices have better gain due to the shorter channel length, thinner gate oxide, and lower threshold voltages), explains the speed improvement of *eight to ten times* that of metal-gate CMOS logic.

SPECIFICATIONS FOR THE 54/75HC LOGIC FAMILY

An operating range of 2 to 6 V dc was chosen for the high-speed CMOS devices. The 6-V dc maximum limit must not be exceeded because of transistor drain-to-source punch-through with short channel lengths. Further reductions in channel lengths will bring about a new 2- to 3-V power supply voltage standard for VLSI devices.

High-speed CMOS (HC) devices have specifications as to the maximum allowed dc output current for all outputs as well as for the V_{cc} potential for metal migration, a phenomenon that occurs if dc current limits are exceeded for long periods. This can degrade the reliability of the circuit.

Input logic levels for high-speed CMOS devices are similar to standard CMOS devices, typically 1.0 V and 3.5 V, using a 5-V dc power supply. Over the complete supply voltage range, the worst-case V_{ih} and V_{il} input levels are typically 70% and 20% of V_{cc}, respectively. In addition, to facilitate interfacing to TTL and TTL-compatible circuits, a subfamily of devices (designated MM54HCT/MM74HCT) are available which operate using TTL input logic levels. See Appendix A for a complete listing of 54/74HC and MM54/74HCT logic devices and CMOS computer devices.

COMPARING OUTPUT CURRENT SINK AND SOURCE

The high-speed CMOS devices (74HC) can drive capacitive loads with the same response time as that of the bipolar devices (75SL). The reason for this can be seen by looking at the comparisons of the output current sink and source capabilities of these logic families shown in Figures 5.9 and 5.10. Here the 74HC CMOS output stages can be seen to compare favorably with those of the 74LS bipolar logic family. It is these large output currents that provide the ability to rapidly drive load capacitance. The 74HC devices require the same attention as the 74LS to PC board layouts and power supply bypass capacitors.

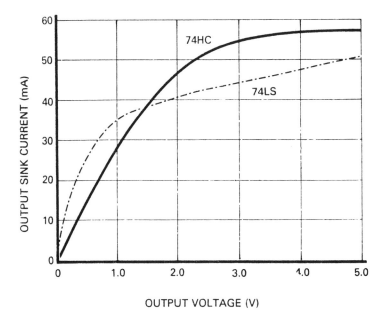

Figure 5.9 Comparing the output sink current capabilities between 74HC and 74LS logic circuits. (Courtesy of National Semiconductor.)

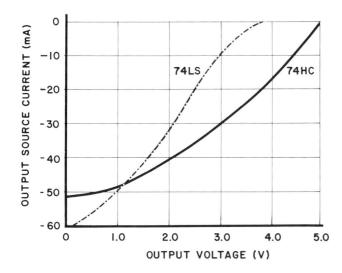

Figure 5.10 Comparing the output source current capabilities of the 74HC and 74LS logic circuits. (Courtesy of National Semiconductor.)

CMOS AND BIPOLAR DESIGN

Bipolar designers who have never used CMOS must remember one important rule: *Never leave an input pin floating.* It must be tied off to V_{cc}, ground, or an output. If left floating, this input can charge up (due to leakage current flow) to the threshold voltage level and turn on both the N-channel and P-channel transistors at the same time. This will cause the IC to draw continuous high and potentially destructive currents from V_{cc} to ground. This problem is aggravated by the large current capability of the 74HC logic family.

CMOS DATA CONVERTERS SIMPLIFY SYSTEM DESIGN

For data conversion using digital-to-analog converters (DACs) and analog-to-digital converters (A/Ds), CMOS offers many advantages. The first is the low-cost high-performance analog switch that is easily achieved: the CMOS transmission gate. These switches are useful in DACs and have allowed a breakthrough in the design of A/Ds.

Analog-to-Digital Converters

The comparator is the key to an A/D converter. It must be fast, with a fraction of an LSB voltage overdrive; exhibit no hysteresis and no oscillations in the linear region; and have high noise immunity.

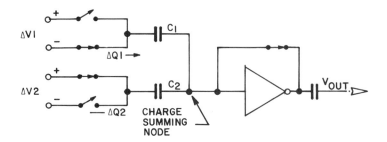

Sampled-Data Comparator

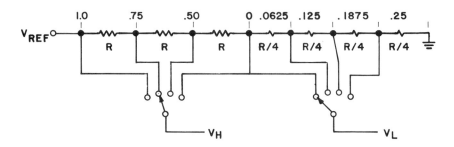

Differential D/A (DDAC) Concept

Figure 5.11 Sampled-data comparator circuit. (Courtesy of National Semiconductor.)

A CMOS sampled data comparator has solved these performance problems and also provides many additional features: for example, true differential analog voltage inputs, useful reference voltage options, and even incorporation of the analog multiplexer within the A/D converter.

To realize the comparator function, a cascade of capacitor-coupled CMOS logic inverters and switches are used as shown in the top portion of Figure 5.11. The differential input voltages are converted to weighted input charges by scaling the value used for each input capacitor (C1 and C2). These input charges are balanced at the input summing node ($^\Delta Q1 = {}^\Delta Q2$ at balance or $^\Delta V1 \times C1 = {}^\Delta V2 \times C2$).

The differential input voltage feature of the sampled data comparator allows us to use a resistor ladder network where zero is not at the bottom of the ladder, as shown in the lower portion of Figure 5.11. Sharing the ladder makes use of all of the tap voltages that are provided by the DDAC a second time to produce 4 additional bits of resolution. The proper reduction

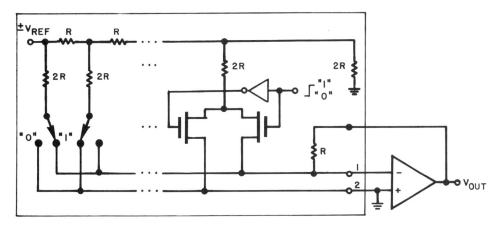

Figure 5.12 Basic CMOS MDAC circuit. (Courtesy of National Semiconductor.)

in the significance of these last 4 bits is achieved by scaling down the value of the input capacitor for these 4 LSBs by a factor of 16.

Digital-to-Analog Converters

The switches available in the CMOS processes allow the realization of low-power-drain DACs, as shown in Figure 5.12. This multiplying DAC (MDAC) uses an R-2R resistor ladder and N-channel current-mode switches. Silicon chromium thin-film resistors are used for the ladder, so the reference voltage (V_{ref}) for the DAC can be of either polarity since there are no parasitic diodes associated with thin-film resistors.

These DACs appear like a memory location to a microprocessor. Decoding the address bus provides the chip select (CS) signal and the digital code on the data bus is read by the MDAC when the write strobe (WR) falls. For the 10-bit and 12-bit MDACs, options are provided which follow data to be accepted in one write cycle or within two bytes.

A complete listing of National Semiconductor's CMOS semiconductor products appears in Appendix C.

CHAPTER REVIEW QUESTIONS

1. Which process requires fewer masks, the bipolar or PMOS?
2. Which semiconductor contains a pentavalent impurity?
3. What is the Miller effect?
4. How does the PMOS process reduce gate capacitance?
5. What contaminant produces problems with N-channel semiconductors?

6. What is the meaning of DRAM?

7. What two advantages does a CMOS-processed IC offer?

8. Describe the configuration of PNP-NPN complementary transistor.

9. What logic function will be performed by the transistor configuration of Figure 5.8?

10. Define the term "micron" and discuss its relationship to decimal measurements.

11. What benefit is realized when integrated circuits can be operated at lower voltage levels?

12. Which has lower interconnect resistance: heavily doped poly or metal?

13. What condition exists if a CMOS IC is said to have "latched up"?

14. What would happen if a negative voltage were applied to a CMOS input pin?

15. What added benefits are achieved by reducing the dimensions of a logic inverter from 125 μm to 64 μm?

GLOSSARY

A/D. Analog-to-digital converter.

Bipolar. Having two poles; a transistor with collector and emitter having identical P or N semiconductor composition.

Chip. A semiconductor substrate on which electronic circuits are produced by microcircuit techniques.

CMOS. Complementary metal-oxide semiconductor.

CS. Clear to send.

DAC. Digital-to-analog converter.

Doped. The addition of impurities to a semiconductor material to increase its conductivity of electrons.

Drain. A terminal of an N-channel field-effect transistor normally connected to V_{cc} or a positive voltage source.

DRAM. Dynamic read-only memory.

Epilayer. A film of semiconductor material deposited on the surface of an integrated chip.

Geometry. The shape, configuration, design, appearance, or layout of an object, transistor, or integrated circuit.

IC. Integrated circuit.

Logic 0. A logic state represented by a dc voltage ranging from 0 V to 1 V.

Logic 1. A logic state represented by a dc voltage ranging from 3 V to 5 V.

Logic family. A series of different logic functions produced by the same fabrication process. For example, the CMOS fabrication process can be used for logic AND, OR, NAND, or NOR gates, inverters, converters, RAM, ROM, and microprocessors; each device becomes a member of the CMOS family.

LSB. Least significant bit.

M2CMOS. Two-metal-layer complementary metal-oxide semiconductor.

Mils. Milliamperes; one thousandth of an ampere.

MOS. Metal-oxide semiconductor.

MOSFET. Metal-oxide semiconductor field-effect transistor.

MSI. Medium-scale integration.

N channel. A semiconductor material that has been doped with a pentavalent impurity which, by covalent bonding with silicon, contains free electrons; a "donor" semiconductor, one that readily donates free electrons.

NMOS. N-channel metal-oxide semiconductor.

Oxide. A material that has been combined with oxygen to alter its original characteristics (e.g., aluminum, a conducting metal, upon oxidation to aluminum oxide becomes a nonconducting insulating compound).

P channel. A semiconductor material that has been doped with trivalent impurity which, by covalent bonding with silicon, contains an available "hole"; an "accepter" semiconductor, one that readily accepts free electrons.

PMOS. P-channel metal-oxide semiconductor.

P2CMOS. Two-polysilicate-layer complementary metal-oxide semiconductor.

Source. A terminal of an N-channel field-effect transistor normally connected to V_{ss} or ground; source of electron movement through the drain terminal to V_{cc}.

SSI. Small-scale integration.

VLSI. Very large-scale integration.

6

Control Devices, Interfaces, and Machine Computers

Progress follows in single file behind the man willing to stick his neck out.
Unknown

The word *control* is perhaps one of the most confusing words in the English language, for it has a minimum of 13 definitions—19 when it is used as an adjective or adverb. However, we can simplify its meaning by narrowing its use to three categories:

1. Manual control
2. Semiautomatic control
3. Programmed control

To give a simple example of how each category of control would function, suppose that you are reading the evening paper as the sun begins to set. You reach up and turn ON a lamp to offset the lowering light level and continue reading without straining your eyes. This would be considered voluntary manual control of the light level, with your eyes serving as light sensors. The actual turning ON or OFF of the lamp would constitute manual control.

If we add an outdoor light-sensitive photocell as a switch to the lamp, similar to those shown in Figure 6.1, we can set the lamp to turn ON automatically when the daylight falls to a given level. You must turn OFF the lamp when you retire for the night or the lamp will burn until morning daylight reaches the sensor turn-OFF set point. If you manually turn OFF the lamp at night, you must remember to turn it ON the next morning to resume its semiautomatic function.

Going a step further, you could replace the manual and semiautomatic methods with a rotating-dial clock timer equipped with an internal switch

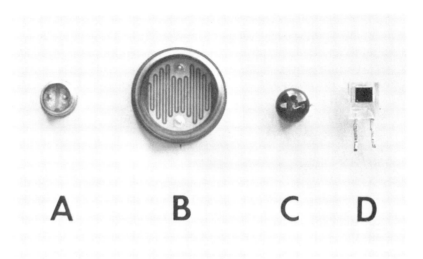

Figure 6.1   A and B photo resistive cells, C photo-transistor, D photo diode.

operated by adjustable ON–OFF tabs which are used to set the ON and OFF time periods. The rotating-dial clock timer, shown in Figure 6.2, has a dual 12-hour dial: one for daylight hours and the other for evening hours. This feature allows you to preset any ON–OFF cycle in a 24-hour period.

The clock timer in Figure 6.2 has only two adjustable tabs and can perform only one ON–OFF time cycle in any 24-hour period. However, there are clock timers that do not have the rotating dial but do have sufficient ad-

Figure 6.2   24-hour clock timer with ON–OFF tabs for setting timed event.

justable taps to preset ON–OFF time-cycle sequences every 30 minutes. The lack of a rotating dial limits the preset time ON–OFF cycles from midnight to noon or noon to midnight and will repeat the preset cycles every 12 hours.

The foregoing examples of simple control functions, although very elementary, do provide a first step toward understanding, leading to the more complex methods associated with computer control and the various types of external devices necessary to accomplish control functions.

CONTROLLABLE EXTERNAL DEVICES

For every manual operation of a machine or process to be replaced by computer control, a specialized external device must be used to perform the human functions of observation; measurement; control of movement and depth; and regulation of speed, pressure, temperature, and flow of liquids or gases. The control devices must not only perform their specialized tasks, but they must also have the capability of relaying precise information as to what it is doing to the controlling computer. We have made remarkable progress in perfecting the speed and accuracy superiority of computer-controlled devices over human control. However, we must still rely on the human qualities of reasoning and decision making.

BASIC OPERATION REQUIREMENTS

All controlled devices must be able to operate and return status information within the current and voltage limitations of the controlling computer's I/O port. Most TTL I/O output circuits will produce a binary logic 1 with an output voltage ranging between 3.5 and 5 V dc at a current between 20 and 40 mA. In cases where the controlled device requires a higher voltage or current to function, transistor interfacing is used to provide the higher device power and to isolate it from the lower voltage or current control signals. Figure 6.3 illustrates two methods by which a 5-V relay may be operated from either a 5- or a 12-V dc source and controlled within the limits of the computer's I/O port. The relay, shown in circuits A and B of Figure 6.3, has a coil impedance of 70 Ω and draws 72 mA to close its contacts. Under normal operating conditions, the relay will pull in between 3.5 and 5 V.

In circuit A of the figure, when a logic $\emptyset$ voltage is applied to the base of the transistor, the base will not be forward biased in respect to the emitter (see the graph in Figure 6.4). In this state, no current will flow through the transistor or relay coil, and the full source voltage will appear at the collector. The resultant +5 V dc at the collector provides a logic 1 feedback to signify that the relay is open.

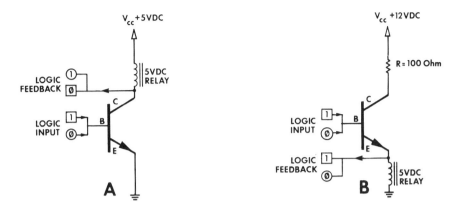

Figure 6.3   Transistor logic driver circuits.

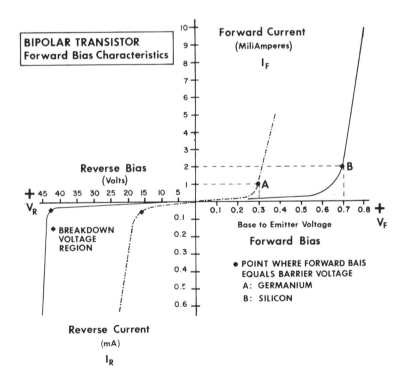

Figure 6.4   Bipolar transistor forward bias characteristics.

Upon receiving a logic 1 on the base, the transistor will be forward biased to conduction, which energizes the relay coil to close its contacts. When this takes place, the full source voltage is dropped across the relay coil and the transistor's collector voltage drops to zero. This produces a logic 0 feedback to indicate the closed state of the relay.

It should be noted that circuit A can be used only if the V_{cc} or source voltage does not exceed 5 V dc. If the source voltage is 12 V dc or greater, the unbiased state of the transistor's base will cause the collector voltage to rise to the source level and exceed the 3.5 to 5-V logic 1 voltage level. Should this happen, the higher voltage feedback would destroy the I/O input circuits.

It is possible to operate a 5-V relay from a 12-V V_{cc} or source by adding a series-dropping resistor to the circuit. The series resistor may be inserted between the collector and the source or between the relay and the transistor's emitter when the collector ties directly to the source voltage. The relay and its series resistor cannot be installed in the circuit A configuration because when the transistor is zero biased, the full source voltage will appear on the collector. In this case, a feedback from the collector would exceed the logic 1 voltage. The best method would be as shown in configuration B, with status feedback taken from the emitter of the transistor.

The operation of circuit B is similar to that of circuit A in that a logic $\emptyset$ applied to the base of the transistor prevents current flow through the circuit and subsequently the full source voltage appears at the collector. With the 5-V relay connected to the emitter and a 100-Ω resistor connected between the collector and V_{cc}, a logic 1 applied to the base would forward bias the transistor to conduction and cause a current to flow through the circuit. When this takes place, the current through the relay coil will develop a voltage of 5 V across the coil while the current flow through the collector's series resistor produces a voltage drop to effectively form a voltage-dividing network which enables the 5-V relay to function at its normal voltage and current rating.

The only apparent problem with circuit B lies in taking a feedback signal from the emitter circuit (the top of the relay coil) because the feedback would be a logic 1 instead of a logic $\emptyset$. In the majority of cases, the feedback calls for a logic $\emptyset$ to indicate a closed state of a relay or switch. In this case, a simple inverter circuit in the feedback path would return the feedback logic state to the normal logic $\emptyset$ to the I/O port.

In configuration A of Figure 6.3, the impedance of the relay coil limits the transistor's current to prevent the source voltage from being pulled down to zero. When a 5-V relay coil is used in combination with a series dropping resistor, with the resistor connected between the coil and the emitter or between the collector and the source, the combined impedance of the resistor and the relay coil limits the transistor's current to prevent the source voltage from being pulled down to zero.

The value of the series dropping resistor is determined by Ohm's law using the voltage difference between the relay coil voltage and the source voltage divided by the relay coil's current rating:

$$R = \frac{E_s - E_c}{I_c}$$

where R = resistance, ohms
$\quad E_s$ = source voltage (12 to 35 V dc)
$\quad E_c$ = voltage rating of relay coil
$\quad I_c$ = rated current of relay coil

Using a 12-V source and 5-V relay as an example, the value of the series dropping resistor can be determined as follows:

$$R = \frac{12 - 5}{0.072} = \frac{7}{0.072} = 97.222 \ \Omega$$

This is a nonstandard value and may be replaced by a 100-Ω 1% tolerance standard resistor. Although there is some leeway in resistor tolerance, the total IR drop (resistance $\times$ current) should not exceed an E_s of 4.5 V.

CONTROL AND MEASURING DEVICES

The control of any function or device that is continuously variable, such as temperature, rate of flow, speed of rotation, depth of drilling, pressure in any form, X-Y axial movement, or viscosity, requires conversion from its analog state to digital form, either binary or binary-coded decimal, before it can be utilized by the computer. The device's analog output must be amplified to a high level for transmission between the device and the computer's analog-to-digital converter. Depending on distance, the signal must be transmitted, and because of subsequent transmission losses, additional amplification may be required before A/D conversion can be performed (refer to Chapter 4).

One method for increasing analog signal levels for transmission is by the Octapak (Figure 6.5) and the Action Pak signal processing module (Figure 6.6).* The Action Pak modules are inserted into the Octapak on-board sockets for each of the Octapak's eight input channels for analog process measurements and control. Pak modules are available for:

1. Limit alarm relays
2. Transmitters/signal conditioners for ac, dc, and frequency analog sensors

*Octapak, Action Pak, and Transpak are registered trademarks of Action Instruments Co. Inc., San Diego, California.

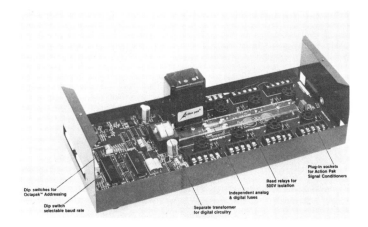

Figure 6.5 Octapak transmission and signal processing unit. (Courtesy of Action Instruments Co. Inc.)

3. Frequency integrators/totalizers
4. Power/reference supplies
5. Math and process functions
6. Ramp generators
7. Process control and drivers

Figure 6.6 Action Pak amplifier and signal processing module. (Courtesy of Action Instruments Co. Inc.)

With an appropriate Action Pak module, the Octapak can condition and amplify analog signals over the range 0 to 10 V on each of its eight analog input channels. The number of input channels may be increased by adding more Octapak units. Each additional unit will increase the number of input channels by 8 to a maximum of 128 channels. The Octapak's analog transmission range varies with the computer's serial I/O port; 100 ft for an RS-232-C port and 4000 ft for an RS-422 port.

Temperature-Measuring Devices

Temperature is one of the most important control or measuring functions to any industrial process. It can also be an important variable parameter in other measurements where temperature has an effect on the measurement's accuracy. In an age where we have the capability of measuring contamination in parts per billion or measuring the size of microelectronic circuit components in millionths of an inch, it is surprising that temperature continues to be measured in quarters of a degree or more. Figure 6.7 illustrates a typi-

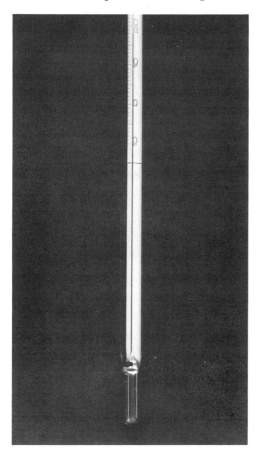

Figure 6.7 Mercury bulb thermometer.

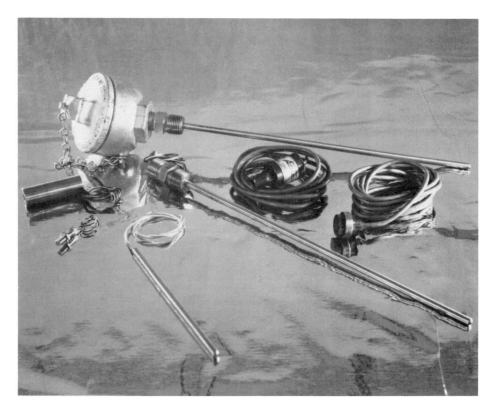

Figure 6.8 Industrial thermocouple devices.

cal laboratory-accuracy mercury bulb thermometer calibrated in degrees Fahrenheit (°F) and Celsius (°C). It takes an exceptional good eye to estimate, with any degree of accuracy, any fraction of a degree on either scale. Mercury bulb thermometer accuracy can be improved by viewing the temperature scale with a magnifying lens containing a reticle graduated in steps of 0.1 mm. In this manner temperature can be estimated to 0.1 °F or °C.

Industrial process temperatures are higher than the 150°C (300°F) range of mercury bulb thermometers and require a different method for high-temperature measurements. The thermocouple provides temperature measurements with a high degree of accuracy within the temperature range in which it is designed to operate. Thermocouple temperature ranges generally run from 0 to 500°C, 0 to 2000°C, 0 to 250°F, and 0 to 2500°F. Figure 6.8 illustrates some of the thermocouple configurations used for industrial process measurements.

A thermocouple consists of two dissimilar metals, in the form of wires, with the dissimilar wires fused together in the form of a bead. The principle of a thermocouple is that when the opposite end junctions are maintained at different temperatures, an electromotive force proportional to the tempera-

TABLE 6.1   Analog Resolution for the Action Pak
Thermocouple

Model number	Range	Resolution (deg)
AP4100-173	0–250°C	0.06
AP4100-1392	0–500°C	0.12
AP4100-1242	0–1000°C	0.25
	0–1000°F	0.25
AP4100-108	0–2000°F	0.5

ture difference is induced. The electromotive force or voltage output of a thermocouple is very small, on the order of millivolts. The thermocouple's ouput voltage is measured and compared to a calibration chart to find the temperature. Each thermocouple has a calibrated chart of voltage versus temperature for the different metal alloys used to make the thermocouple currently in use.

Thermocouples for the higher-temperature measurements, 500 to 2000°F or °C, are generally made with platinum wire joined to a platinum-rhodium alloy wire to withstand the temperatures without melting. Thermocouples, when properly made and with the end junctions stabilized at 0°C, are capable of fractional-degree accuracy which must be maintained during transmission and signal processing to the analog-to-digital converter. The degree of analog resolution obtainable by the analog signal processing of an Action Pak thermocouple module is shown in Table 6.1.

In cases where very high precise measurements are required at temperatures under 100°C (212°F), a temperature-dependent resistor called a *thermistor* can be used. The resistance of the thermistor has a negative coefficient which causes the resistance to decrease with increases in temperature. Accuracies on the order of 0.01°C and 0.03°F can be obtained with thermistor devices similar to the one shown in Figure 6.9. This thermistor is only 0.040 in. in diameter, sealed into a 0.0625-in. glass tube.

Precision measurements can be made at higher temperatures with the use of another type of temperature-dependent resistance called *RTD*. Most RTD devices are constructed of platinum wire to withstand temperatures above 1000°C or 2500°F. The positive coefficient of platinum results in an increase in resistance which is directly proportional to increases in temperature. RTD temperature measurements can be made with an accuracy of 0.03°C or 0.05°F below 500° and to 0.1° above 1000°. RTD transmission and signal processing is accomplished by a Transpak (see Figure 6.10), which uss a 20-mA current loop circuit to accomplish signal transmission to an Octapak system or directly to the control computer's RS-232 or RS422 serial I/O 20-mA current loop inputs.

Temperature-measuring devices can be used in conjunction with other measuring devices where the function being measured imposes a number of temperature-related variables. In the following example, a firm required mea-

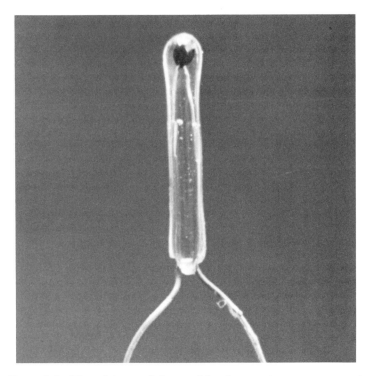

Figure 6.9   Thermistor used for precision temperature measurements.

Figure 6.10   Transpak 20 mA current loop for transmitting RTD analog temperature measurements. (Courtesy of Action Instruments Co. Inc.)

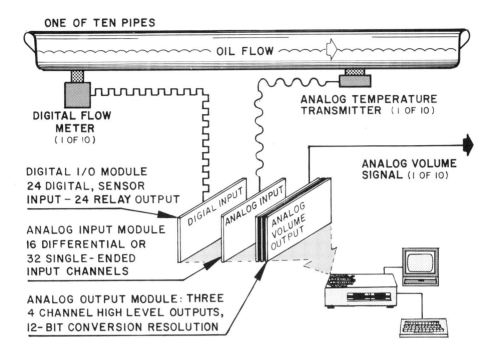

MEASUREMENT OF OIL FLOW VS TEMPERATURE FOR CONVERSION TO TOTAL MASS

Figure 6.11   Method of measuring oil mass. (Courtesy of Action Instruments Co. Inc.)

surements of oil flowing through 10 different pipes for computing mass from volume. To accomplish this measurement, 10 frequency inputs were attached to the 10 pipes to measure flow (a volumetric reading), together with 10 analog inputs representing temperature (thermocouples). The 10 frequency transducer inputs were fed into a digital input/output interface and the 10 analog thermocouple inputs were fed into an analog-to-digital converter.

Using the computer's integral timer and counters, the frequencies of the flow signals are determined and stored as a variable. Using the flow data, temperature data, and a table of flowmeter characteristics, the computer dynamically compensates for the viscosity of oil in each pipe and provides an analog output for each pipe that is directly proportional to the total mass delivered from each pipe. Figure 6.11 illustrates the digital flowmeter and temperature transmitter paths to the computer's digital and analog interface modules as well as the analog high-level output module.

Controlling Process Temperature

Having achieved accuracies of 0.03° and better in measuring temperature, industry turned toward controlling temperature to the same degree of accuracy. A common method of temperature control familiar to everyone is the household thermostat, with its ability to control the temperature climate of our homes.

The principle of the thermostat, like that of the thermocouple, is based on the use of two dissimilar metals. In the case of the thermostat, the dissimilar metals are laminated together into a small bar which has the ability to bend or bow with changes in ambient temperature. This is due to the differences in the coefficients of expansion between the dissimilar metals. Having this property, the bimetallic bar can be used to make or break the electrical circuits controlling a furnace or air conditioner using the ambient temperature of the house. A typical thermostat for home heating and air conditioning is shown in Figure 6.12. The dual power contacts, one for heat control and the other for cooling control, are located in the glass tube below the coiled spring at the upper left of the thermostat.

The thermostat is set at the temperature we wish to have maintained by moving to the desired temperature the pointer on a calibrated temperature dial. When this is done, the dial turns a cam which releases electrical circuit contacts to apply power to the heating or air-conditioning system. The heating or cooling function is set by the switch shown at the bottom left of the figure. As the ambient temperature rises to the set point, the bimetallic

Figure 6.12 Typical bimetallic strip thermostat.

strip bends away from the electrical contacts and breaks the circuit to turn off the heat source. For air conditioning, the reverse is followed. The set point of the thermostat is set lower than the ambient temperature to turn on the air conditioner. When the ambient temperature drops to the desired set pont, the bimetallic strip curves away from the electrical circuit contacts to the air conditioner and turns off the cooling process. The thermostat represents a very simple method of temperature control; however, its control leaves much to be desired for industrial process control because of the inherent overshoot and undershoot characteristics of the bimetallic strip.

Overshoot and Undershoot Characteristics

When a thermostat is in the heating mode, the bimetallic strip set point is generally the ambient temperature at which we wish to have the heating unit turn off. However, when the heating unit cuts off, the plenum is still very hot and the system's blower fan will continue to blow hot air into the house. This causes the ambient temperature to rise above the original set point. This is called *overshoot* because the temperature overshoots the thermostat's set point. There will be a time lapse, based on the rate of heat loss of the home, before the ambient temperature falls to the thermostat's set point.

The bimetallic strip has remained curved away from the heating system's electrical control contacts because the ambient temperature has been above the set point of the thermostat. Now comes a cooling period during which the bimetallic strip must reverse its curve to again make electrical contact with the heating circuit's electrical system and resume heating.

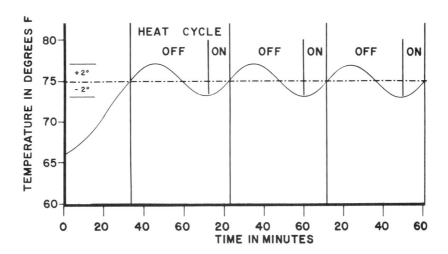

Figure 6.13 Overshoot-undershoot characteristics using thermostat heat control.

Figure 6.13 graphically illustrates the overshoot and undershoot characteristics of a home thermostat that is in a fully insulated home that has aluminum heat-loss factor. The temperature variation of +2 to −2° is well within the comfort zone for people. The heating cycle, ON 10 minutes–OFF 40 minutes, would be more economical than equal ON–OFF periods because the heating or cooling motors draw maximum current until operating rpm has been reached. Consequently, the more starts of the system's motor per hour, the higher the operating cost.

CONTROL TRANSDUCERS

Industrial machines and processes frequently use hydraulic and pneumatic pressure to operate devices where the controlled movement requires force of as little as 1 ounce per square inch (psi) to several thousand psi. Obviously, pressure must be measured in a way that does not produce a pressure loss and the measuring device must be constructed strong enough to withstand the pressure under measurement safely. Figure 6.14 illustrates the ruggedness of the housing for a pressure transducer.

A *transducer* is a device designed to receive information in the form of a physical property and to convert it to an electrical signal that is proportional to the magnitude of the physical form. The input energy may be acoustical, electrical, optical, mechanical, or pressure. The converted

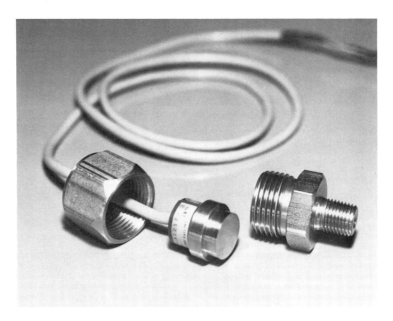

Figure 6.14   Pressure transducer and housing.

output may be in the form of analog or digital signals representing the input information.

A simple example whereby mechanical movement can be converted to an electrical signal would be the action of a turntable's stylus and a crystal or magnetic cartridge. As the turntable rotates a record, the stylus rides in very small spiral grooves which have varying side wall dimensions that represent acoustical information in the form of music, speech, or noise. The varying widths of the record's grooves cause the stylus to move or vibrate in proportion to the groove's physical shape. In turn, the stylus mechanical movement or vibration is transformed to a variable-inductance movement within an inductive field to induce an electrical signal containing all the mechanical frequencies represented in the record's grooves. The ability to convert mechanical movement into an electrical signal puts the magnetic cartridge and stylus into the category of transducer.

Technically speaking, any device that has the ability to convert electrical signals into mechanical motion is a transducer even though we may not recognize it as such. Not many people would look upon an audio speaker as a transducer, yet it converts electrical signals into mechanical movement of the speaker cone to produce sounds that we can hear. A television receiver not only converts electrical signals into sound, but also converts electrical signals into optical signals that we can see. Similarly, few would classify an electric motor as a form of transducer even though it converts electrical current into mechanical movement.

Figure 6.15 shows a very valuable type of motor for control applications. The stepper motor differs from the conventional electric motor in that its armature shaft rotates in precise degree steps instead of rotating con-

Figure 6.15 Haydon Big Inch stepper motor and 180:1 reducing gear train. (Courtesy of Haydon Switch and Instrument.)

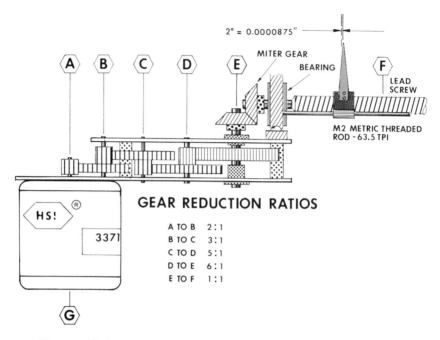

Figure 6.16 Stepper motor and gear train application to achieve precision measurements. (Courtesy of Haydon Switch and Instrument.)

tinuously. A second difference is that the stepper motor runs by a sequence of pulses instead of by continuous ac or dc current. In addition, the direction of a stepper motor's armature rotation can be reversed by varying the sequence of pulses to its dual field windings. The Haydon Big Inch stepper motor, shown in Figure 6.15, is available with a choice of operating voltages, degree steps, and unipolar or bipolar operation (see Appendix D).

Figure 6.16 illustrates how a reduction gear train can be used to greatly increase the accuracy of measurements or how position is to be controlled by a stepper motor. The Haydon Big Inch stepper motor's gear train uses a variety of gear ratios to achieve an overall reduction ratio of 180 : 1. For each 360° revolution of the stepper motor gear (A), the gear train output shaft rotates only 2°. The miter gears on this shaft and the metric lead screw serve only to couple the two together and to permit the lead screw to operate at right angles to the gear train's output gear shaft. Miter gears would permit independent vertical or horizontal measurements or positioning to the gear train.

As a further example of how a measurement would be performed with this stepper motor/gear train configuration, we will assume that the gear motor operates in steps of 15 angular degrees. This would require 24 pulse sequences to accomplish 360° rotation of stepper motor gear A. The 180 : 1

reduction of the gear train would rotate the M2 metric lead screw only 2°. The lead screw, having 63.5 threads per inch, would move its traveling pointer 0.015748 in. in one 360° revolution. In this case, the lead screw has rotated just 2°, and by dividing the movement of one 360° rotation by 180 (the reduction ratio), the total movement of the pointer would be 0.0000875 in. or 0.2203 μm. If we obtained the stepper motor driving pulses from an on-board real-time clock, only 24 s would have been required to make a measurement to this degree of accuracy. In determining the measurement, the only variable is time, as reduction and threads per inch are constants. The measurement needs only the total pulses divided by 24 and when multiplied by 0.0000875, the final measurement is obtained. Similarly, the pulse rate may be 24, 240, or even 2400 pulses per second without changing the result.

Stepper motors, due to their small size and driving currents, generally have very low holding torque. This is defined as the torque required to deflect the motor one full step. The force of the motor's torque is given in millinewtons per meter (mN·m) or in ounce-inch-second (oz-in.-s). By coupling a stepper motor to a reduction gear train, the motor's running torque is effectively increased by the reduction ratio of the gear train, negating gear and friction losses.

CONTROL AND MEASURING DEVICE COMPUTER INTERFACES

There are only two ways in which data can be moved into and out of a computer: through direct connections to the computer's address and data bus lines, common to internal circuits such as the ROM, RAM, and EPROM internal memory circuits, or by an input/output port like the serial RS-232 or parallel I/O ports. The reader may wonder why, if the computer has both serial and parallel I/O ports, additional interfaces are required for control and measurement. An 8-bit microprocessor (MPU) has the capability of addressing 256 I/O ports. However, the general RS-232 computer I/O has only three lines to the MPU: serial input, serial output, and common ground. Similarly, the RS-232 I/O has only five lines going to its output: serial out, serial in, 20-mA current loop in, 20-mA current loop out, and common ground. In practice, both serial and 20-mA I/Os provide a choice of only two methods for input and output from the serial I/O and do not necessarily indicate two different serial I/Os. For parallel I/O operation, a separate bidirectional parallel I/O is required for multiple parallel operation.

An examination of the typical MPU memory map shown in Figure 6.17 will disclose that there are a number of I/O addresses that can be addressed. Therefore, by adding additional digital interface modules such as the one shown in Figure 6.18, an on-board RS-232 I/O can be used to address 72 digital outputs, and with additional digital modules, this can be expanded to a

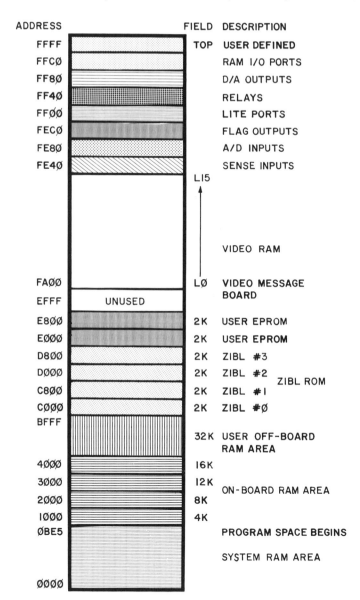

ADDRESS

	FIELD	DESCRIPTION
FFFF	TOP	USER DEFINED
FFCØ		RAM I/O PORTS
FF8Ø		D/A OUTPUTS
FF4Ø		RELAYS
FFØØ		LITE PORTS
FECØ		FLAG OUTPUTS
FE8Ø		A/D INPUTS
FE4Ø		SENSE INPUTS

L15

↑

VIDEO RAM

	LØ	VIDEO MESSAGE BOARD
FAØØ		
EFFF	UNUSED	
E8ØØ	2K	USER EPROM
EØØØ	2K	USER EPROM
D8ØØ	2K	ZIBL #3
DØØØ	2K	ZIBL #2
C8ØØ	2K	ZIBL #1
CØØØ	2K	ZIBL #Ø

ZIBL ROM

BFFF	32K	USER OFF-BOARD RAM AREA
4ØØØ	16K	
3ØØØ	12K	ON-BOARD RAM AREA
2ØØØ	8K	
1ØØØ	4K	
ØBE5		PROGRAM SPACE BEGINS
		SYSTEM RAM AREA
ØØØØ		

ZILOG Z80° MPU MEMORY MAP

Figure 6.17   Zilog Z80 MPU memory map chart. (Courtesy of Action Instruments Co. Inc.)

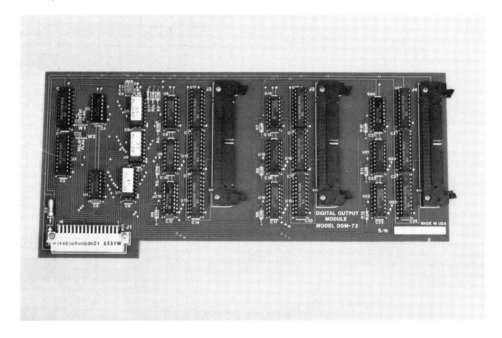

Figure 6.18 Digital interface module for 72 additional RS-232 I/O ports. (Courtesy of Action Instruments Co. Inc.)

total of 432 digital outputs. The memory map also shows reserved addresses for A/D inputs and D/A outputs for analog data. With the appropriate analog I/O interface shown in Figure 6.19, the computer can address 32 single-ended or 16 differential inputs. Like the digital inputs, the analog inputs can be expanded with additional interface modules to a maximum of 256 analog inputs.

The memory map also shows the addresses reserved for sense inputs and flag outputs. Thus, with a digital I/O module, as shown in Figure 6.20, 25 sense inputs and 24 flag outputs may be addressed. These can also be expanded to a maximum of 48 inputs and outputs by the addition of another interface module.

There are a number of special interface modules that can be used with the reserved addresses for relays and lites remaining on the memory map. You can add solid-state relay modules to add up to 24 relays at voltages from 90 to 140 V ac, 180 to 280 V ac, 4 to 16 V dc, or 10 to 32 V dc with a current capability of 3 A for each voltage. There are interface modules for thermocouples and RTD temperature-measuring devices which can expand the number of devices in use to 150 thermocouples or RTD units per module. The interface for strain gauges and other low-level analog sensors can handle up to 40 inputs with a 12-bit plus-sign conversion resolution.

Figure 6.19 Analog interface module for 32 single-ended or 16 differential additional analog input ports. (Courtesy of Action Instruments Co. Inc.)

Figure 6.20 Digital sense and flag interface module to add 24 input/output lines. (Courtesy of Action Instruments Co. Inc.)

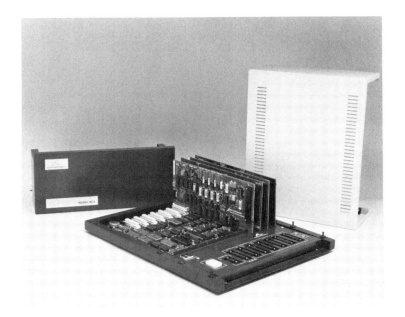

Figure 6.21   BC3 industrial computer showing its interface module expansion capabilities. (Courtesy of Action Instruments Co. Inc.)

Judging from the size of the interface modules shown in the figures, the reader may wonder how it is possible to add so many modules to any computer in view of the fact that most computers have so little available space that the most they can accommodate is eight add-on interfaces (and most have no more than five). Generally speaking, most industrial control applications seldom require more than three or four modules at a time. However, Action Instruments, Inc.'s BC3 industrial control computer has the capability of utilizing 10 interface modules at the same time with its on-board Actibus connectors. Figure 6.21 presents the BC3 industrial computer with its housing removed so that you can see how the interface modules are installed. The figure shows four modules in place and six white sockets available for additional modules.

MACHINE SITE COMPUTER INSTALLATIONS

As pointed out in Chapter 2, locating computer peripherals in an industrial area can generate a variety of data problems from machine or other electrical interference. Signal transmitter and conditioning devices, which must be located in proximity to the measuring device, are generally installed in moisture-proof metal cases with the signal-connecting links run through metal pipe. Figure 6.22 shows a Transpak 20-mA current loop transmitter in an ex-

Figure 6.22 Transpak mounted in an industrial explosion proof housing. (Courtesy of Action Instruments Co. Inc.)

plosion-proof housing. Threaded pipe is attached to the end fittings to provide signal link shielding and protection from possible explosions in volatile atmospheres.

The use of control computers in the proximity of the machine or process is becoming commonplace in a number of industries. Figure 6.23 shows a BC3 control computer in an industrial process area. It shows a number of visible hazards to the computer:

Figure 6.23 Example of a machine computer installation for industrial control. (Courtesy of Action Instruments Co. Inc.)

Figure 6.24 Example of a machine computer installation. (Courtesy of Action Instruments Co. Inc.)

1. The computer table is within inches of the stairs, obstructing passage up and down, and is a hazard to the computer.

2. Permanently installed electrical equipment is located too close to the BC3 mainframe housing.

3. The computer has not been permanently mounted to the table and could easily be damaged by accidental bumping of the table, causing the computer to fall to the floor.

4. The sign on the wire fence cautions that the computer is in an "ear-protection" area, indicating a high noise level. There is no evidence of acoustical or dust protection for the computer.

A better industrial machine site computer operation is shown in Figure 6.24. Here a BC3 computer has been permanently installed in a ventilated protective housing in an obviously very clean environment. Its location with respect to the instruments it controls or monitors does not obstruct access to the instruments or the computer. Notice that the CRT monitor is positioned at the same eye level as the instrument dial on the right. The missing keyboard and data storage peripherals suggest that this computer is operating under the control of a host computer.

CHAPTER REVIEW QUESTIONS

1. Why would it be necessary to install a light-sensitive switch outdoors instead of on an an interior lamp and activated by the ambient light level of the room?

2. What eight human control functions must be accomplished by controlled devices for a computer to control a machine operation?

3. What is the voltage range for a binary logic 1?

4. What is the normal pull-in voltage of a 5-V dc relay?

5. Define the term *forward bias* as it applies to an NPN bipolar transistor.

6. When a 5-V dc relay is connected between a transistor's emitter and ground, what happens to the emitter voltage when the transistor conducts current?

7. What prevents the source voltage from being pulled down to zero if a relay coil is connected between the source voltage and a transistor's collector and the collector voltage drops to zero?

8. In calculating a series resistor to enable a 5-V dc relay to operate at higher voltages, which would be preferred if the calculated resistance is not a standard value: a resistor slightly higher or lower than the calculated value?

9. List three functions that the Octapak can perform when it has the appropriate Action Pak modules installed.

10. What are the average maximum temperature scales of a mercury bulb thermometer?

11. What are the dissimilar metals used in a thermocouple to measure temperatures above 1000°?

12. What effect would temperature have on a resistance that has a positive temperature coefficient?

13. Which motor operation would be the most economical: one that runs continuously or one operating in an equal ON–OFF sequence?

14. What function does a transducer perform?

15. How does a stepper motor differ from a conventional electric motor?

16. How many pulses would a stepper motor require to make one complete rotation if its step were 7.5°?

17. What is the maximum number of I/Os that can be addressed by a microprocessor?

18. What can be used to increase the number of I/O lines in a computer?

19. What are the two ways in which a microprocessor addresses any data location?

GLOSSARY

Ambient. Surrounding or circulating about; prevailing temperature of surrounding air.

Control. A device for regulating and guiding a machine or process; computer as the source of control for a device, machine, or process.

I/O port. A bidirectional electronic circuit by which data can be entered or received from a computer; serial or parallel I/O ports.

Parameter. A determining factor, characteristic; a variable term in a function; a mathematical constant or a variable in a parametric equation.

Plenum. A section of a home heating system where the heated air pressure is greater than the atmosphere pressure; a major heated air distribution duct.

Reticle. A design or scale printed or scribed on a glass insert in the focus point of a lens used for viewing objects at an enlarged or magnified size.

RTD. Resistor temperature dependent. A device used for high-temperature measurements in which resistance is directly proportional to changes in temperature.

Stepper motor. An electrical motor that functions in specific degree of rotation steps as opposed to continuous rotation at a given number of revolutions per minute.

Thermistor. A negative coefficient resistance that lowers its resistance in proportion to increases in temperature.

Thermocouple. A high-temperature measuring device made of dissimilar metal wires; when their end junctions are maintained at constant but different temperatures, a current will flow in proportion to the difference between the two temperatures.

Thermostat. A device used to control heating and cooling by means of a bimetallic strip of dissimilar metals which bends to make or break electrical contacts when the ambient temperature reaches the set point of the thermostat.

Transducer. A device that will accept energy in one form and transmit the input energy in another form (e.g., input electrical energy and transmit mechanical energy).

TTL. Transistor-to-transistor logic.

7

Ultraprecision-Controlled Devices and Intelligent Systems

It is said that there are four types of engineers: those who watch things happen, those who wait for things to happen, those who wonder what happened, and the one who makes things happen.
Unknown

Microelectronic circuits ushered in a new era of ultraprecision measurement and positioning requirements. The microinch and micron units of measurement superseded the standard of one to two thousands of an inch for dimensional tolerances. The practice of slicing microcircuit wafers requires positioning the wafer to an accuracy impossible to obtain without the aid of the computer. Beams of wafer-slicing lasers must be focused and aligned to a fraction of a micron to realize maximum productivity at minimum waste.

Friction is an important parameter when associated with ultraprecision positioning devices. Friction, produced from a sliding movement of a positioning device, will generate sufficient heat to create thermal expansion in metal components of the device. The degree to which thermal expansion can affect the accuracy is dependent on the temperature coefficient of expansion of the type of metal used to fabricate the device. One method to reduce or eliminate friction is to isolate the positioning device with a cushion of air called *air bearings*. The air bearing provides friction-free movement of the positioning device and simultaneously provides thermal isolation to heat. Figure 7.1 shows a slide-type ultraprecision positioner that employs the air bearing principle. The slide surrounds a heavy steel supporting frame to position loads weighing up to 60 lb. Its horizontal movement can be controlled by computer-compatible, smooth-motion drives such as stepper motors, linear dc motors, or lead screws. Typical position accuracies of 4 microns for a movement of 2 in. (10 μm over the 5-in. span of the device). Figure 7.2 shows the inside surface of the slide and the compressed air ports that pro-

Figure 7.1   Micro Robotics Center, Inc.'s air bearing slide positioner. (Courtesy of Micro Robotics Center, Inc.)

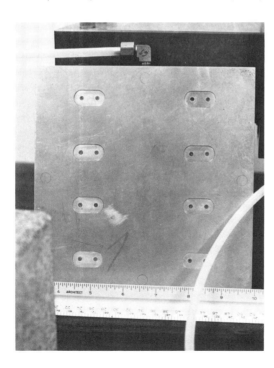

Figure 7.2   Air port side of the air bearing slide. (Courtesy of Micro Robotics Center, Inc.)

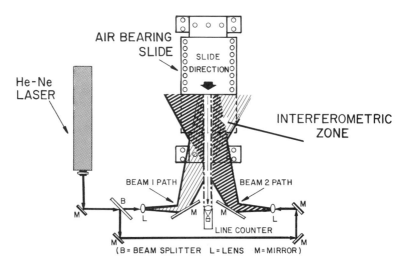

Figure 7.3 Laser interferometer set-up for interferometric measuring. (Courtesy of Micro Robotics Center, Inc.)

duce the air bearing. Remote position measurements of the slide can be made with a laser interferometer by optically counting the interferometer's pattern of lines as the slide moves toward the laser beams. Figure 7.3 is a diagram of a type of interferometric configuration as it might apply to a slide-type positioner. In this example, a laser beam is directed through a beam splitter to produce two laser beams. One beam passes through the beam splitter to a lens which causes the small-diameter beam to expand in proportion to the focal length of the lens. The expanding beam is then reflected by a mirror onto the moving slide as a reference beam. The portion of the laser beam reflected by the beam splitter is directed by mirrors over a longer path to its expanding lens than that of the first beam. The difference in beam path lengths produces a small phase shift between the frequencies of the two beams. When both beams are reflected off the end of the slide, the phase difference produces an interference pattern in the form of lines or fringes. The greater the phase difference, the more lines or fringes will be produced. Figure 7.4 presents a block diagram of a *micromation* system by which a computer can control the movement of a micromechanical device and measure its travel by means of interference patterns which can be optically translated into data representing the position of the device to an accuracy of 5 to 10 μin.

Traditionally, the *micrometer* has been the standard of precision measurements when accuracies were on the order of 0.001 in. By coupling the micrometer to motors and a computer, the same basic micrometer can measure to 1 micron or less. Figure 7.5 shows three types of motor-operated micrometers and a conventional micrometer. Table 7.1 compares the performance of the motorized micrometers. From the table it can be seen that the

MICROMATION SYSTEM

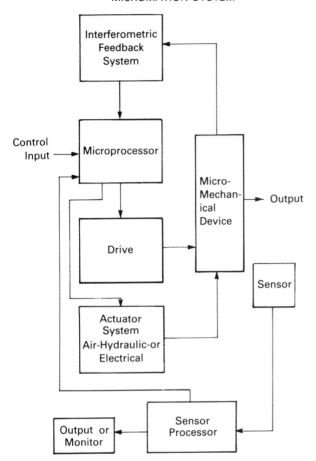

Figure 7.4 Block diagram of a micromation system. (Courtesy of Micro Robotics Center, Inc.)

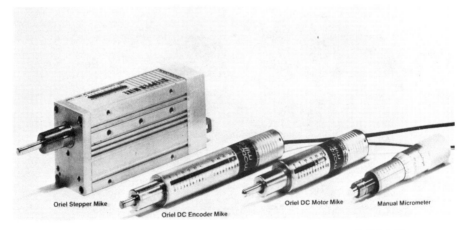

Figure 7.5 Oriel motor driven micrometers. (Courtesy of Oriel Corporation.)

TABLE 7.1   Performance Comparison of Motorized Micrometers

Type	Resolution (μm)	Speed (mm/min)	Load (kg)	Position readout (μm)
Stepper motor				
Oriel Stepper Mike	1.0	0–60	14	1.0
Dc motor				
Oriel Motor Mike	0.02	0.03–12	1	No
Oriel Encoder Mike	0.02	0.03–12	1	0.1

stepper micrometer has 5 times the speed and 14 times the load capability of the dc motorized micrometers. Its disadvantages are the vibration produced with every step and the fixed 1 μm per step, which limits its resolution. Position readout is accomplished by the computer counting the drive pulses of the stepper motor.

The dc motor micrometer, Motor Mike, has very good resolution, is virtually vibration free, and is the most economical way to motorize the micrometer. However, it has no means for a computer to display a position readout. By adding an optical shaft encoder to the dc motor, a method is provided by which the computer can read pulses from the encoder directly to provide a position readout. Optical encoders are the key factor for computer control in closed-loop applications.

Working with lasers or optics frequently calls for ultraprecision positioning of lens, pinholds, beam splitters, or mirrors within a fraction of a

Figure 7.6   Oriel three axis lens rotor. (Courtesy of Oriel Corporation.)

degree per second of arc. Figure 7.6 shows a three-axis lens rotator with ultraprecision circular and angular capability. Adjustments about the optical axis may be varied between 3 arc min/min to 18 deg/min with a resolution of 0.12 arc sec. Adjustments about the orthogonal axes are made to 0.14 arc sec. The reference to the orthogonal axes means that adjustments to any one axis does not noticeably affect the adjustment of another axis. In addition to the optical axis, the three-axis rotator also provides vertical and horizontal axis adjustments by means of two encoder micrometers, and all three axes adjustments provide position readouts through an LED display.

CONTROL BY LIGHT AND COMPUTER VISION

In the early days of factory automation, one of the first approaches to emulate human vision was the use of photocells and a light source spanning a conveyor belt to count the number of items passing along the belt. When an object passed through the light beam, the light on the photocell would be momentarily blocked, which causes an electrical pulse to energize a small solenoid that increments a mechanical counter by one digit. The photocell was next adapted to sorting different-size items on a conveyor belt by size. By setting a number of photocells and lights at different heights, each photocell would respond only to items of a given size and would not respond to smaller sized items. Once an item of the desired size darkens the photocell, its electrical pulse triggers its corresponding ejection solenoid and the part is pushed off the conveyor belt into the appropriate hopper or bin. In this manner, objects would be sorted by ejecting the tallest part first, followed by smaller sizes according to the variations of object heights and the number of photocells/ejectors utilized. Figure 7.7 illustrates the photocell and companion light source for counting objects passing through the light beam and for sorting items by height or size. In addition, a typical circuit provides a method by which when the light is cut off by an object, the photocell's resistance increase can turn on a bipolar transistor and close a relay between the emitter and common ground. In this instance, the photocell's light resistance is approximately 100 Ω, which causes its series resistor, R1, to produce an IR drop equal to 90% of V_{cc}. This will forward bias the base slightly above the emitter voltage and produce a small current flow through the transistor but not enough to energize the relay coil. When the photocell is blocked by an object, the photocell's dark resistance rises rapidly to a point where the series resistor's IR voltage drop is approximately 5 V. This puts a forward bias of 7 V on the transistor's base to drive the transistor into full conduction and closes the relay in the emitter circuit. Since the collector is tied directly to V_{cc}, the relay coil in the emitter and R1 prevent V_{cc} from being pulled down to zero.

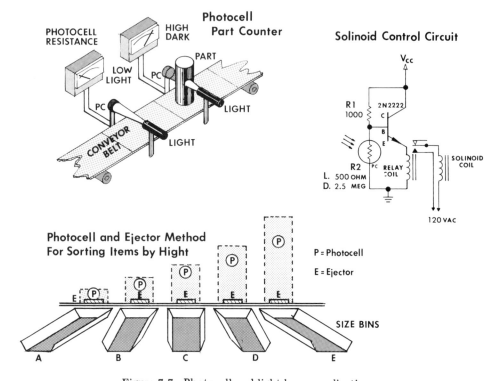

Figure 7.7 Photocell and light beam applications.

MACHINE AND COMPUTER VISION

Photodetector systems perform economically in automated systems without the ability to detect visual characteristics of the production items. Their disadvantages are the number of detector/light sources required for size sorting by height, limitations in diameter or thickness detection, and the ability to identify objects by shape. The state of the art in video equipment has made possible a new method for sorting, inspecting, and measuring production items that has made the photocell method obsolete. While commercial video systems can be used for a number of tasks, the speed at which objects can be processed depends on the complexity of the object or how many features must be identified. Selection of a commercial video system is governed by the desired processing speed, resolution of the image, and the contrast of the image.

The Video Camera

Computer vision systems look at parts through a conventional television camera. Each picture is divided into a number of picture elements called *pixels,* an acronym for *picture elements.* A typical video camera has an array of 320 × 240 pixels, the aspect ratio of the *vidicon* picture format, with each pixel capable of rendering 16 *gray levels*. If the entire array information is to be processed and stored in memory, the minimum number of memory elements required would be 1,228,800 bits or 153,600 bytes plus the x-y locations of each pixel. The computer then compares the information in memory with software data of the object's key features to identify the object.

From the standpoint of memory requirements, it would be most inefficient to use the video information as it is received from the video camera because the majority of the camera video signal pertains to an area in which little or no image information is present. The object's features which are evaluated by the computer are primarily edges, which constitute a very small percentage of the total picture. By compressing the image data from the video camera so that only the information related to edges is stored, the processing speed and RAM memory space requirements for a large number of patterns will be reduced. This technique is called *running-length encoding.* Analysis of the video data, compressed in this manner, takes place concurrently with another process that actually recognizes the pattern of the image.

The following example illustrates how the running-length encoding can reduce memory storage requirements. When video *recognition* of handwritten signatures is performed, an analysis of the signatures reveals that only 15% of the pixels contain gray or black in the image area. Running-length encoding enabled a reduction of image data by a factor of 3 and reduced the gray-scale level from 16 to 4—needed to show a clear image of a signature. In this manner, the data storage for a signature is reduced from 7200 bytes to 300 to 600 bytes. Figure 7.8 shows the logo of Applied Intelligent Systems, Inc. as it would appear photographed in four gray levels with a resolution of 128 × 128 pixels.

The processing speed of running-length encoding processing varies with the pixel area of the image. A small image area, on the order of 128 × 128 pixels, requires approximately 0.1 s to process as compared to the process time of 0.8 s for an area of 320 × 240 pixels.

Pattern recognition in most commercial video systems takes place through a software algorithm called *connectivity analysis.* This procedure reduces a binary image to its connected components by classifying samples according to statistical considerations. By this manner, connectivity analysis uses the statistics of a scene to gauge if the object is good or bad. The system is "trained" by running a number of "good" objects past the camera to enable the image analyzer to accumulate statistics of the features that characterize the object. The system then uses the accumulated statistics to check objects under inspection to accept or reject the object. Allowable tolerances or deviations can be set automatically or manually to reduce sensitivity to certain defects.

Connectivity analysis can identify geometric characteristics of an object. These include area, maximum width, height, perimeter, centroid posi-

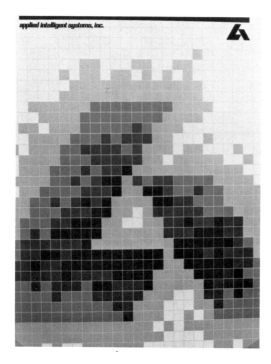

applied intelligent systems, inc.

Figure 7.8 Video display from a 128 by 128 pixel view and in 4 gray scales. (Courtesy of Octek, Inc.)

tion, orientation, elongation, and compaction index. Elongation is the ratio of major-axis and minor-axis second movements, and the compression index is the ratio of the square of the perimeter to the area.

Some system software subroutines have user calls to make these calculations, depending on the object features to be analyzed. Where additional geometric data is required, special programs can be written and combined with those producing geometric characteristics analysis.

The speed of a vision system is inversely proportional to the amount of information being processed. Large numbers of objects can be processed quickly—3600 objects per minute is the upper limit—or can be further limited by the scanning speed of the camera. At this speed, the system generally can report only the presence or absence of an object. The more slowly the objects move past the camera, the more information the system can calculate about what it sees. For an example, area, position, number of holes, or orientation can be calculated in approximately 0.25 s, which is a rate of 240 objects per minute in a 128 $\times$ 128 pixel area. If the camera view is expanded to a 256 $\times$ 256 pixel area, the process time increases to 1.0 s, a rate of 60 objects per minute.

The practical resolution of an area-scan *machine vision* system is currently 320 $\times$ 480 pixels. Where applications call for greater definition, the object must be scanned in sections with multiple cameras or through a step-and-repeat process. A step-and-repeat scanner can divide the object into sections scanned one by one. In this instance, processing speed is proportional to the number of pixels and object positioning time.

Objects moving on a conveyor belt may be scanned at right angles to the movement of the belt on the horizontal or vertical axis by means of line-scan cameras. These cameras are available with resolutions up to 2048 pixels. Higher resolution may be obtained by staggering line-scan cameras on either side of the conveyor belt. However, two line-scan cameras with a 2048-pixel area would each have a throughput 200 times slower than a 320×240 pixel camera. Resolution can be enhanced by viewing one edge of the object and referencing the opposite edge to a known coordinate.

Image contrast is also a process speed factor. The contrast of the object against its background must be greater than the localized lighting variations around the feature of interest. Lighting variations are caused by point light sources and interference from ambient light. Object features that the system must extract, such as edges or holes, must be distinguished from the local background by 15 to 25% of the overall intensity range. Using a scale of 1 to 10, the system can distinguish an edge if the background intensity is at level 3 and the edge is illuminated at a level of at least 4.5. Normal edge-level lighting should be a level 5.5 or greater.

Rear lighting is preferred because it produces greater image contrast. Frontal lighting can be used where surface features must be extracted. However, the lighting intensity must be sufficient to swamp interference from ambient lighting of the object.

Computer Vision Requirements

Figure 7.9 illustrates the integrated architecture of the Octek, Inc.'s computer vision system. Two video cameras are interfaced to the 8-300 data bus via two RS-170 video ports by software control of the desired port selection. The architecture facilitates synchronization from the composite video, camera sync signals, or signals generated from an internal hardware clock. It is capable of a resolution of 320×250 pixels from 16 gray levels or 320×480 pixels from four gray levels. The digitizing rate is 60 fields per second from either vidicon or charge-coupled-devices (CCD) cameras. An optional function control keyboard allows interactive control by soft-programming the keys for the desired function control and enters into the 8×300 data bus via the digital I/O port.

The four RS-330 video ports allow the configuration of a number of simultaneous displays: a 16-level gray scale and an eight-color display; four monchrome or two four-color displays, acting independently of each other; or a color display of 16 software-selected colors from a choice of 64 color selections. The displays also have the capability of moving a block or cross-hair cursor freely over the digitized image on the monitor without overwriting or interfering with the image.

Computer Vision
Integrated Architecture

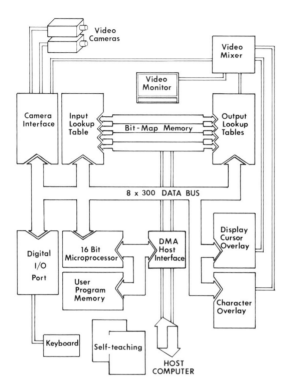

Figure 7.9   Computer Vision integrated architecture diagram. (Courtesy of Octek, Inc.)

The heart of the system is a Signetics 8 × 300 16-bit microprocessor using an 8-bit peripherals bus for efficient execution of commands. It takes 325 ns to fetch a complete instruction cycle from memory, manipulate it, and return it to memory or to an I/O port, all while video continues digitizing in real time. The multiport memory provides concurrent read, write, and DMA operations for processing digitized video images while maintaining image memory integrity and uninterrupted screen refresh. The multiple lookup tables allow preprocessing and postprocessing data for development of interactive programs. A third lookup table is used to create the color video output.

Video Image Capabilities

Visualizing the capabilities of machine vision is difficult because most people cannot conceive the effects produced in digitizing an image. We are prone to expect commercial television images instead of a blob of different shades of gray blocks such as the logo shown in Figure 7.8. A better example can be seen in Octek, Inc.'s Content Verification System (CVR), used to verify con-

Figure 7.10   Octek Content Verification System—equipment photo.
(Courtesy of Octek, Inc.)

tainer presence and its contents, as shown in Figure 7.10. Although the setup
in Figure 7.10 is only a demonstration, it can perform the same visual inspection with packages moving along a conveyor belt. Figure 7.11 shows the digitized image that will display on a monitor. In this example, the image is a
circular bottle and shows that it contains the desired contents, as indicated

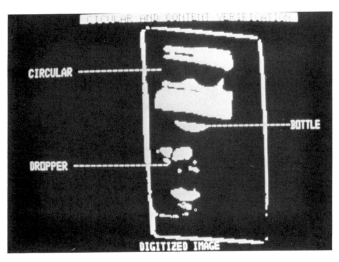

Figure 7.11   Binary image of package shape and content verification.
(Courtesy of Octek, Inc.)

Figure 7.12 Octek Target Application System, "TASK." (Courtesy of Octek, Inc.)

by the legend "Dropper." Figure 7.12 shows a complete Octek system for label verification, inspection of electronic printed circuit boards, dimensional tolerances of cast or extruded parts, and robot control and sensing. This is a Target Application System, called TASK, designed for stand-alone factory floor operation. Figure 7.13 shows the Octek 20/20 Vision Development System (VDS), designed as an automation tool to solve the machine vision task.

Figure 7.13 Octek 20/20 Vision Development System (VDS). (Courtesy of Octek, Inc.)

Figure 7.14 320 by 250 pixel video display of visual inspection of a good automobile engine's piston. (Courtesy of Octek, Inc.)

Suppose that we examine some of the VDS images to get a better understanding of what machine vision is capable of doing when it is displayed in high-resolution images. Figure 7.14 is an inspection image of a good automobile piston. The image contrast and resolution shows the quality that can be obtained with 320 × 240 pixels. However, it still lacks commercial television contrast and resolution. Figures 7.15 and 7.16 show production stains and plating defects along the side walls of the piston. Keep in mind that

Figure 7.15 Production piston showing side sidewall stains. (Courtesy of Octek, Inc.)

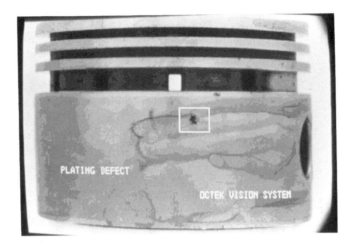

Figure 7.16 Production piston showing plating defects. (Courtesy of Octek, Inc.)

these three figures have not been digitized to produce binary images. The image in Figure 7.17 has been digitized to a binary image to emphasize the plating defect shown in Figure 7.16. Note that the binary image contains only two shades: white and black. The *binary* image in Figure 7.18 is a reversal of the binary black-and-white areas. A grinding error is shown in Figure 7.19 and has been digitized and displayed in four gray levels.

To go the extra step in showing what can be done with the appropriate

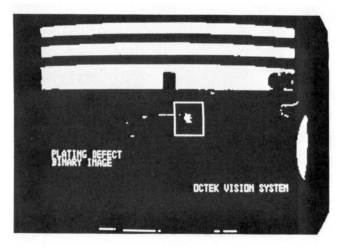

Figure 7.17 Binary image of the piston with plating defects. (Courtesy of Octek, Inc.)

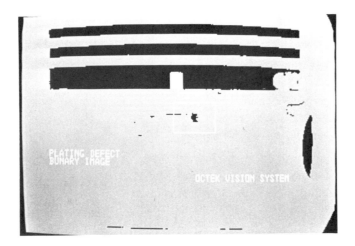

Figure 7.18   Brunary image of piston with plating defects. (Courtesy of Octek, Inc.)

software and machine vision, Figures 7.20 and 7.21 show the machine vision inspection of printed circuit boards and how "OK" is stamped on good circuits and an error code on bad. Figure 7.20 indicates that all pads are soldered except the last pad on the bottom left. The error code is directly above the pad. In Figure 7.21, all pads along the bottom are OK, but the two upper right-hand pads have error codes. The extreme right top pad contains no solder and the pad next to it has a lead off center.

Figure 7.19   Production piston with grinding defects digitized and in four gray level display. (Courtesy of Octek, Inc.)

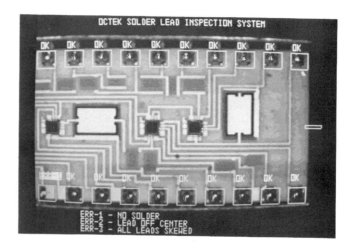

Figure 7.20 Video inspection of printed circuit board. (Courtesy of Octek, Inc.)

Machine vision systems are expensive, ranging in cost from $18,000 to over $100,000. Even with the most expensive system, the accuracy and speed for complex inspection readily justifies the initial expense. Take a computer keyboard assembly and inspection as an example. The manufacturer's incoming inspection performs a 10% inspection of the incoming key caps to weed out defects in character quality and plastic molding. Once the

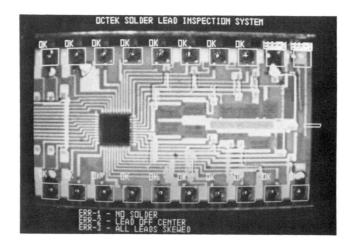

Figure 7.21 Video inspection of printed circuit board. (Courtesy of Octek, Inc.)

key caps have been assembled onto the keyboards, they are inspected visually with the aid of plastic overlays that point out bad characters, incorrectly attached caps, misaligned caps, and other physical defects. Human visual inspection is far from perfect because by the time the completed keyboards reach the equipment manufacturer, the rejection rate runs as high as 12%.

Machine vision inspection of completed keyboards examines a 60-key keyboard in 15 s or one key in 0.21 s and reliably detects 99.9% of key cap defects or assembly errors. In addition, it labels no more than 1 good key cap in 10,000 as an error. Thus to reduce assembly and inspection errors from 12% to 0.01% is ample justification for machine vision inspection cost by any manufacturer.

Applied Intelligent Systems uses a lower-cost method for image inspection, based on key geometric features to distinguish specific objects with their Pixie-1000 Machine Vision System and a GE-220 camera using a field of 128 × 128 pixels. The visual inspection digitation process used by the Pixie-1000 is the PIX-4 (Pixie encoding algorithm), called cellular automata-based image processing, accomplished through image algebra. When presented with a problem, Pixie literally asks itself a series of questions, each designed to describe the desired object in progressively finer detail.

For example, suppose that Pixie is looking for 6-32 machine screws,

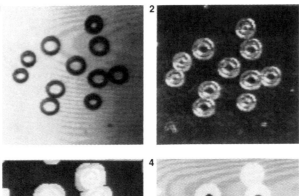

1. **First frame:** The PIXIE camera records a frame of 13 washers, 4 of which are undersized. The washers are non-uniformly illuminated and rest on a dirty conveyor belt. Some are touching.

2. **12 milliseconds:** PIXIE has identified the outline of each washer and has suppressed the background to a uniform black.

3. **84 milliseconds:** PIXIE has visually "filled in" the hole in each washer. A single pixel dot remains in the center of each correctly sized washer. (The hole-size fault tolerance is programmable.)

4. **129 milliseconds from the first camera frame:** The identification is complete. Black spots locate the centers of correctly sized washers. A microcomputer will now transfer the coordinates to the materials handling equipment.

Figure 7.22   Applied Intelligent Systems's Pixie-1000 image identification inspection process. (Courtesy of Applied Intelligent Systems.)

Figure 7.23 Fairchild CCD camera option for the Pixie-1000 Machine Vision System. (Courtesy of Fairchild Camera and Instrument Corp.)

which are $\frac{1}{4}$ in. long. It might test each of the image objects in its viewing field by applying the following questions:

1. Does the object have a silhouette $\frac{1}{4}$ in. long?
2. Is it $\frac{1}{8}$ in. thick?

Figure 7.24 Charge Coupled Device (CCD) used in the Fairchild CCD-4001 Process Camera. (Courtesy of Fairchild Camera and Instrument Corp.)

3. Does it have a head?
4. Is the head $\frac{1}{4}$ in. wide?
5. Does it have threads with the correct pitch?

Pixie will then mark those locations in the viewing field where the answer to all questions is "yes" and will report their locations. Figure 7.22 illustrates the four Pixie image steps in the process to identify and locate specific objects by size where two or more sizes appear in the image. Note that images 1 and 4 show the effects of RF interference. These can be eliminated with the optional Fairchild CCD-4000 series 256 $\times$ 256 pixel camera and RF filtering shown in Figure 7.23 and its CCD image target, shown in Figure 7.24. See Appendix C for Fairchild CCD-4001 Process Camera specifications.

ARTIFICIAL INTELLIGENCE

The more human functions a computer can perform, the closer it comes to acquiring artificial intelligence. We have already given the computer the human capability of vision, and now we have developed a means whereby a computer can recognize human speech and actually snthesize a vocal reply. The Votan VTR 6000 Voice Terminal, shown in Figure 7.25, transforms a computer to respond to continuous speaker-dependent recognition (CSDR) and speaker-independent recognition (SIR), which enables the computer to understand verbal instructions and return a verbal response.

Speech Technology Fundamentals

There are two basic fundamentals to speech technology: *speech recognition*, where the computer accepts and understands verbal input using speaker-dependent recognition (SDR), or *speaker-independent recognition* (SIR) and

Figure 7.25   Votan VTR 6000 Voice Terminal. (Courtesy of Votan.)

voice output (VO). VO is the reverse of speech recognition; the computer talks by using speech compression encoded from an actual human source. Speech compression is divided into two categories: *voice response* (VR) and *voice store-and-forward* (VSF).

Speaker-dependent recognition recognizes previously trained words for the user's library of word templates. The voice terminal's training is accomplished by speaking the intended words into a microphone several times. This enables the computer to digitize the words and record the voice patterns. Multiple-user word libraries can be stored on individual disks to prevent computer error when different users must share the same Votan voice terminal. To activate the voice terminal, the user's vocabulary disk is brought on-line and the computer then matches the pattern of each spoken word with the pattern stored in the user's private vocabulary. With the continuous speech recognition function, the user can speak without pauses or breaks between individual words. The Votan voice terminal is capable of accuracy levels greater than 99%.

The speaker-independent recognition mode has the ability to recognize, accept, and respond to a fixed set of words spoken by a wide range of speakers without prior training and development of individual vocabularies. Recognition in this case is "independent" of the speaker because the vocabulary is predetermined by the manufacturer and usually consists of the numbers zero through nine, yes/no, and other simple words. The SIR systems are programmed to accept English inputs from a broad cross section of people with regional accents and variations of speech.

The Votan voice terminal can be used in security applications where speaker verification (SV) determines "are-you-who-you-say-you-are" through voice-print analysis. In this way, SV is "verbal fingerprinting," but faster and friendlier. Thus, by simply speaking a name or password into a microphone, the vocal input is transformed into digital information which also includes certain unique spectral voice qualities of the speaker. It is these qualities that differentiate people speaking the same word and enables the computer to prevent unauthorized access. On of the essential functions in word recognition or speaker verification modes is "dynamic programming pattern matching." This function allows for variations in the way a person may say the same word or phrase from day to day.

Voice Output

Most artificial speech-producing computer interfaces produce speech through speech synthesis and speech encoding. The Votan voice terminal uses a technique called "digital speech compression" to encode speech from a human source. These speech compression methods are sophisticated proprietary algorithms based on digital spectral transforms that reduce memory requirements without lessening vocal quality. Speech compression "strobes" voice

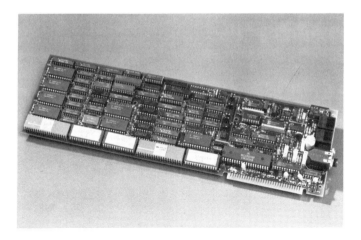

Figure 7.26   Votan VPC 2000 Voice interface for personal computers. (Courtesy of Votan.)

inputs by capturing a number of samples (bits) per second for future reconstruction; bit rates run from 4800 to 14,400 bits per second. In this manner, the voice retains the richness, identity, emotion, and inflection of the original speaker.

Voice response (VS) and voice store-and-forward (VSF) are types of voice output using speech compression techniques. VR is used with static messages where vocal input is stored in digital form in the system's memory and played back at any time by software control. VSF involves real-time encoding, compression, and storage of speech messages for later retrieval. VSF handles temporarily stored and continuously changing messages, whereas VR is a static system. Figure 7.26 shows the Votan VPC 2000 Voice Card designed to provide voice recognition with compatible personal computers (PCs).

CHAPTER REVIEW QUESTIONS

1. What property of metals causes sliding friction to affect the accuracy of position measurements?

2. List three methods by which an air-bearing slide can be driven to its desired position with ultraprecision accuracy.

3. What is an interferometric image?

4. What is the accuracy of a standard micrometer?

5. Which motor drive of ultraprecision motor-driven micrometers produces the most vibration?

6. The Oriel three-axis lens holder performs angular and circular adjustments. What is the resolution of its angular adjustments in degrees?

7. What were the characteristics of the first electro-optical device used in automated production?

8. What would prevent a bipolar transistor's collector from causing V_{cc} to fall to zero when the transistor is biased to saturation or maximum current drain?

9. What is a pixel?

10. What are the minimum number of gray levels needed to identify a handwritten signature?

11. What is the typical pixel area for a high-resolution machine vision camera?

12. What would be the maximum number of object-presence detections that a machine vision camera could detect negating filed scanning vibrations?

13. How is a vision system computer trained?

14. How does conveyor belt speed affect the amount of information a machine vision system can produce?

15. Without changing conveyor belt speed, what can be done to increase video information from an object?

16. In terms of information, what is throughput?

17. What is the maximum number of usable gray-scale levels found in machine vision systems?

18. What is the field digitation or frames-per-second rate of most video cameras?

19. What is the instruction cycle time of a Signetics 8×300 16-bit microprocessor?

20. What is the function of lookup tables in the Oriel machine vision system?

21. What does the term "dropper" refer to in a content verification system?

22. How many gray scales are contained in a binary image?

23. What is a binary image?

24. How many error codes were shown on the printed circuit inspection illustrations?

25. List three advantages of machine vision inspection that offer possible justification for capital investments to implement such a system.

26. What is a voice print compared to in identification of a person?

27. In a Votan VTR 6000 system's computer vocal response, is the voice synthesized in tonal qualities of the user's voice print?

GLOSSARY

Accuracy. The precision in which measurements are made of length, width, height, diameter, and angular properties of an object; the ratio of words correctly recognized to the number of words spoken. Determines the accuracy of a voice recognition system.

Air bearing. Air, under pressure higher than atmospheric, surrounding a movable device or part in a manner that the air cushion does not permit the device or part to come in contact with friction-producing surfaces.

Algorithm. A programming term defining a step-by-step solution to a given problem.

Arc min. One arc minute is equal to $\frac{1}{60}$ of 1 degree of arc; 1 degree of arc contains 60 minutes.

Arc sec. One arc second is $\frac{1}{60}$ of 1 arc minute; 1 arc second is $\frac{1}{3600}$ of 1 degree.

Beam splitter. An optical device or plate containing a special front surface reflective coating with equal transmission and reflection properties.

Binary image. A black-and-white image stored in memory as 0's and 1's. Images appear as silhouettes on the video monitor.

Bit rate. Number of samples per second in capturing voice data for future reconstruction.

Blob. Any group of connected pixels in a binary image. A generic term including both objects and holes.

CCD. Charge-coupled-device camera. An image sensor using semiconductor arrays which enables the output of one to provide the input stimulus to the next.

Centroid. The midpoint of the x and y axes of an object; a blob feature measurement.

Connectivity. Connectivity analysis, a procedure that analyzes the relationship of various pixels within an image to define separate blobs.

Contrast. The relative difference between black-and-white images.

CSR. Continuous speech recognition; allows user to speak without pauses or breaks.

Cursor. A small distinctive pattern (rectangle, cross hair, or hatch) used to highlight specific regions of an image.

Digital spectral transforms. An algorithm for reducing voice computational requirements and eliminating inaccuracies associated with analog filter systems.

Digitation. The process of converting an analog video image into digital brightness values that are assigned to each pixel (gray scale).

Discrete word recognition. Also known as isolated word recognition, each word must stand alone bounded by a period of silence in order to be recognized by a voice processor.

Edge. A distinguishable change in pixel values between two regions. Defining all pixels above a threshold as an object and all below a threshold as background.

Encoder. A device or counter capable of transforming data in one form to another. An optical shaft encoder can count the rotation speed of a motor by a reference line painted on the rotating motor shaft and convert it to a output signal capable of driving LED digital displays or computer analog input data.

Fabricate. The process by which raw material or completed subsections are put together, assembled, constructed, or processed into a finished form or product.

Features. Simple distinguishable data attributes such pixel amplitudes and edge point locations.

Gray level. A quantitized measurement of pixel brightness or irradiance.

Gray scale. The range of gray levels used in the image; the gradient steps between pure white and pure black.

Image. A replica of a physical object or scene, captured electronically or photographically, which can be reproduced and viewed at a later time.

Image processing. Transforming one image into another which has more desirable properties for analysis.

Interferometer. A device producing two paths of light with one path longer than the other and slightly out of phase, which will produce a wavefront or lines denoting the degree of phase differential produced by distance or contour of the surface or device onto which the two light beams are projected.

Interlace. The image displayed on a monitor or television monitor is displayed 30 times per second, with even-numbered lines displayed for $\frac{1}{60}$ of a second and odd-numbered lines displayed for the next $\frac{1}{60}$ of a second. The retention of vision property of the human eye blends the odd- and even-line images into a higher-resolution image than is possible using noninterlaced vidio images.

Lookup table. Memory that sets input and output values for gray-scale thresholding, window, inversion, and other display or analysis functions.

Machine. Computer perception, based on visual sensory input, to develop a concise description of a scene depicted in an image.

Micrometer. A mechanical C-shaped device with a calibrated spindle used for precision measurement of sheet, bar, or round stock; a measuring device capable of measuring to 0.001 in.

Phase shift. A condition where two identical frequencies arrive at a common point, with one frequency leading or lagging the other from 1 to 180°.

Phonemes. The building blocks and smallest units of sound. They are language specific, with English comprising 60 different phonemes.

Pixel. The individual elements in a digitized image array. The array consisting of 320 horizontal elements and 240 vertical elements.

Raster. Parallel horizontal lines drawn by the electron as it scans across the video monitor at a rate of 525 lines interlaced for commercial television receivers or 420 lines interlaced for video monitors.

RS170. The Electronic Industries Association standard for governing monochrome television studio electrical signals.

RS330. The Electronic Industries Association standard governing closed-circuit television electrical signals.

Run-length encoding. A data compression technique in which the image is raster-scanned and only the lengths of runs of consecutive pixels with the same color (black or white) are stored.

SDR. Speaker-dependent recognition. Recognizes previously trained words from a particular user's library of word templates.

Shading. A method using levels of gray scale to differentiate a specific blob from its neighboring black objects and the white background.

SIR. Speech-independent recognition: ability of a device to recognize, without prior training, a wide range of speakers through the use of a universal word template.

Speech recognition. Computer acceptance and understanding of verbal input using speaker-dependent or speaker-independent recognition.

Speech synthesis. Creating words from sonic chunks called phonemes without human voice imprints or patterns.

SV. Speaker verification: determining the identity of a person through voice-print analysis or vocal fingerprinting.

Temperature coefficient of expansion. The rate of expansion of metals, glasses, woods, or plastics with changes in ambient or applied temperature; the amount of expansion or contraction exhibited by a temperature change of 1 degree Fahrenheit or Celsius.

Template. A digital representation of a vocal input.

Threshold. A selected pixel gray level.

Throughput. The rate at which information is processed through a computer.

Tolerance. The allowable variation, over or under a specified dimension, on a machining blueprint. Fractional measurements given as $\frac{1}{4}, \frac{1}{2}, \frac{3}{4}, \frac{7}{8}, \frac{13}{16}$, and so on, typically have tolerances of plus or minus $\frac{1}{16}$ in. For dimensions given in decimal inches; 0.250, 0.500, 0.750, or 0.875 in. will typically give a tolerance as plus or minus 0.002, plus 0.002 minus 0.000, or plus 0.000 minus 0.003.

Vision. The process of understanding the environment based on sensing the light level or reflectance of objects.

Wafer. A silicon disk on which microcircuit devices are fabricated.

Window. A selected portion (square or rectangle) of an image; a limited range of gray values.

8

Industrial Robots

Build a better mousetrap and the world will beat a path to your door.
Proverb

The generally accepted term for describing the modern industrial computer-controlled machines is the industrial robot. However, these machines are not actually robots in the true sense of the word. The dictionary defines a robot as a machine, resembling a person, with the ability to perform routine tasks on command, the mobility to perform its task at any location, and the ability to accomplish this under its own power. Although industrial robots do have the capability of a number of different movements in a radius around their base, they are not capable of moving from one location to another under their own power. To emphasize this point, disconnect an industrial robot's computer and it reverts to a powered machine needing human guidance control to function.

Creating a machine with the ability to move from one point to another under its own power does not create a true robot because it must move on command and be able to carry out manipulative tasks. The automobile is a machine that has the capability of moving from one location to another under its own power. It does not qualify as a robot because it cannot perform routine tasks or respond to vocal or computer commands. The human appearance noted in the dictionary's description of a robot cannot improve or add to the efficiency of an industrial computer-controlled machine. A better definition of an industrial robot would be a reprogrammable machine designed to handle material or tools in the performance of a variety of tasks, including performing the functions of inspection, measurements, verification of manufactured parts, and the manipulation, fabrication, and assembly of components in the production of a finished product, all under the guidance of a computer's control program.

THE INDUSTRIAL ROBOT

An industrial robot consist of three basic components: the machinery or hardware, the command center or control computer, and the software. Movement of the machine's *manipulator* or arm depends on the task to be performed. Figure 8.1 illustrates the four basic designs governing the geometry of manipulator movements; Cartesian, cylinder, spherical, and articulated. It should be noted that most present industrial robots have the capability of manipulator movements in two to six axes, whereas the basic movement designs in Figure 8.1 show only three axes of movement: the Z axis for manipulator elevation, the X axis for base travel, and the Y axis for reach of the manipulator.

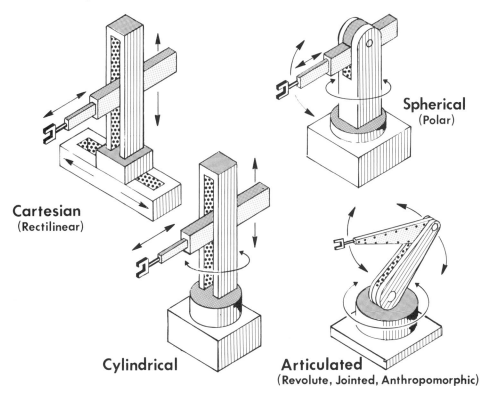

Figure 8.1   Four basic movements of a robots manipulator. (Courtesy of Cincinnati Milacron.)

Basic Design Configurations

The *Cartesian design* enables the machine's manipulator to move across the X axis from one extreme to the other and from the top to the bottom of the Z axis. Manipulator movement on the Y axis is accomplished by means of a telescoping section of the manipulator. The *cylindrical design* retains the Cartesian design Z- and Y-axis movements and replaces the horizontal X axis motion with one that rotates 360°. The *spherical design* retains the cylindrical design X and Y axes and replaces the Z axis with a swivel movement for greater Z-axis movements. The *articulated design* retains the spherical design X and Z axes, then eliminates the telescoping section in the manipulator Y axis by incorporating a second swivel motion at the shoulder to permit the entire manipulator to move along the Y axis.

Yaw, Pitch, and Roll

In order to produce an industrial robot with the maximum application flexibility, three additional movements must be added to the basic design configurations. Figure 8.2 illustrates the yaw, pitch, and roll movements which

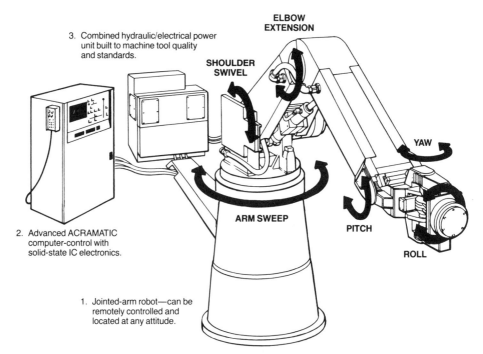

FIGURE 1 - CINCINNATI MILACRON T³ ROBOT SYSTEM

Figure 8.2 End effector yaw, pitch, and roll axes. (Courtesy of Cincinnati Milacron.)

Figure 8.3 Model manipulator
showing +90 degree yaw position.

are installed at the *end effector* or hand section of the manipulator. *Yaw* is
a 180° movement or swivel along the X axis of the base. Figures 8.3, 8.4, and
8.5 show a manipulator in the +90°, 0°, −90° yaw positions. *Pitch* is a 180°
swivel along the Z axis of the manipulator. Figures 8.6 and 8.7 show a manipulator with its X axis at −90° and its Z axis pitch in its +90° and −90°
positions. *Roll* is the circular 360° rotation of the manipulator's end effector
on either X or Z axes. We must not overlook the end effector's gripping action as shown in Figures 8.8 and 8.9, which illustrate open and closed positions of the gripper.

In each of the four basic designs, the end effector gripper was in a fixed
position in respect to the manipulator. Since neither the manipulator nor
end effector could be rotated, only objects that were at right angles to the
gripper's fingers could be handled. The roll axis movement made it possible

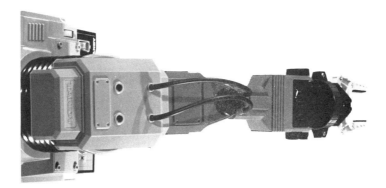

Figure 8.4   Model manipulator showing 0 degree yaw position.

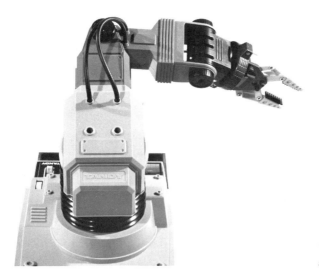

Figure 8.5 Model manipulator showing −90 degree yaw position.

for the gripper to rotate to any position and grip objects regardless of their orientation to the manipulator.

By adding pitch and yaw movements in addition to the roll function, the end effector acquired the flexibility to work at any angle in relation to the position of the manipulator. This increased flexibility of the manipulator and its end effector made it possible for the robot to perform humanlike tasks of assembly and fabrication on an automated production assembly line. Figure 8.10 illustrates the reach flexibility of the six-axis industrial robot.

Figure 8.6 Model manipulator showing +90 degree Z axis position.

Figure 8.7   Model manipulator showing −90 degree Z axis position.

In normal operation the manipulator sweep is typically +120 to −120° horizontally in respect to the shoulder axis and +110 to −70° in the vertical plane of the shoulder axis.

Figure 8.8   Model manipulator shown with gripper in open position.

Figure 8.9   Model manipulator shown with gripper in closed position.

JOINTED ARM - REACH FLEXIBILITY

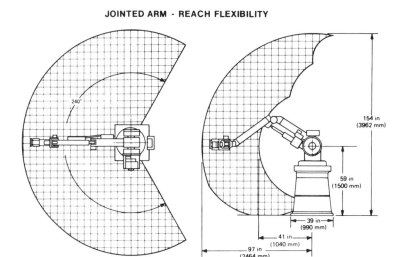

Figure 8.10 Manipulator reach flexability. (Courtesy of Cincinnati Milacron.)

Robot Power Sources

Industrial robots are designed with an operating power source determined by the work to be performed, the weight of objects to be handled, or the speed of the process requirements. Hydraulic pressure powers the robot in cases where the payload varies from heavy to medium weights and the operation requires dexterity with robot resilience. *Resilience* is the robot's ability to "give" when placed under stress, such as hard contact with an immobile surface.

Ac and dc electrical power is used where robots are engaged in tasks that demand a great deal of repeatability or accuracy. In comparison with hydraulic-powered robots, robots powered by electricity typically carry light- to medium-weight payloads. Where fast velocity or light payloads are required, and where environmental considerations prohibit the use of other types of powered robots, the industrial robots are powered by pneumatic pressure (compressed air). Industrial robots are further classified as either point-to-point, nonservo motion, or continuous path/controlled path servo-motion feedback. The point-to-point mode (nonservo) is a control method whereby a series of positions are programmed by a series of electrical or mechanical stop switches. Only stop points can be so programmed since the lack of servo feedback does not permit the computer to determine work position other than the preset stop points.

Servomotion robots are equipped with automatic feedback systems that constantly monitor the robot's motions in order for the computer to exercise precise and flexible control of the robot. Control of this degree is necessary for *continuous-path operation*, where the computer maintains absolute control over the entire path of motion, or *controlled-path operation*, where the computer controls and coordinates all axes of motion along the entire path of motion and enables the robot to move in a straight line between programmed points at a specified velocity.

Control of an Industrial Robot

The control of any automated device depends on the complexity of the device. The simple photocell counter or sorting device required only the manual throwing of a switch lever to power the device. As the controlled device became more complex, a programmable control device became a necessity. This may have been no more than a cam-shaped dial on an electric clock motor, and the shape of the cam controlled the sequence of control. As the industrial robot obtained the capability of repetitive tasks, the nature of its control became equally complex and dictated the use of a computer and appropriate control software to assure accurate and reliable repetition.

Control Computer Requirements

The control computer must have the capability of directing an industrial robot's functions through control software, subroutine programs, and sensory feedback information. A production computer's memory requirements exceed that of a number of microcomputers because its memory capacity must be sufficient to be able to interrupt the software control in process whenever a sensor feedback signals an interruption that calls for a subroutine task before resuming normal operation, and return to the control program once the subroutine program has been completed. Provisions must also exist for the robot machine's operator to interrupt or request changes in the control program according to the on-the-floor conditions in the production area, such as required tool changes, assembly line halted, or material shortages. Most computer-controlled industrial machines provide the machine operator with a keyboard and CRT monitor system for this purpose.

Putting It All Together

The selection of an industrial robot is a compromise of trade-offs versus effectiveness. Axial movement, other than that of the X axis, adds weight to the manipulator, which reduces the manipulator's payload weight capacity. As a general rule, an industrial robot's axial movements must be held to the minimum to realize the maximum efficiency and payload capacity.

Figure 8.11 presents an exploded view of an abbreviated industrial robot configuration showing only key sections in a six-axis design. Each axis

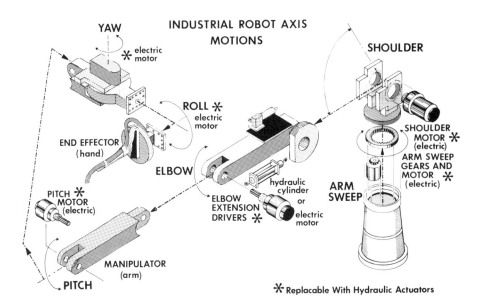

Figure 8.11 Relationship of all sections of a robot manipulator with axle drive motors and axial movements. (Courtesy of Cincinnati Milacron.)

is shown with electric motors providing the driving power. To upgrade these axial functions to a more efficient operation and to heavier duty, the electric motors may be replaced with either a rotary- or a linear-acting hydraulic or a combination of hydraulic and pneumatic actuators.

At first glance it would appear that one need only replace an existing electric motor with another with more horsepower. However, if the manipulator was structurally strong enough to accommodate the increased weight and size of a higher horsepower electric motor, the overall power losses through the motor efficiency and gear train would justify changing to the more efficient hydraulic or combination hydraulic/pneumatic power systems.

Disadvantages of Electric Motor Robot Power

1. Size and weight ratio to torque power
2. Requires gear train for linear motion
3. High starting currents compared to running current
4. Maximum efficiency 80 to 85%
5. Overall motor/gear train efficiencies 70 to 74% (based on 1% loss for each gear meshed)

Hydraulic and pneumatic systems have specialized disadvantages:

ROBOT
End Effector Grippers

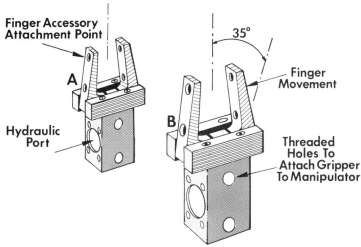

Figure 8.12 Comparison of gripper sizes.

1. Requires air- or fluid-pressurized supply tanks
2. Hydraulic or pneumatic high-pressure plumbing
3. Initial installation more expensive than for electric motor installation

The hydraulic and combination hydraulic/pneumatic actuators offer low weight increases in proportion to increases in output force, low profile for interchangeability with existing equipment, and the ability to function safely in a hazardous or explosive environment. Figure 8.12 shows two hydraulic end-effector grippers to illustrate the weight increase in the end effector resulting from increases in finger pressure and static payload weighs. The gripper in Figure 8.12A has a weight of 3.7 lb and can exert a finger pressure of 305 psi at a static load of 235 lb. The gripper in Figure 8.12B has a weight of 7.2 lb and its finger pressure has been increased to 628 psi at a static load of 625 lb.

INDUSTRIAL ROBOTS ON THE PRODUCTION LINE

The industrial robot first appeared on the automobile assembly lines during the late 1970s as a joint development venture of General Motors Corporation and Unimation, Inc. The original robot has been continuously updated in features and capabilities until it has matured as Puma, shown in Figure 8.13.

Figure 8.13 Unimation Inc.'s
PUMA robot. (Courtesy of
Unimation, Inc.)

Occupying no more space than a human worker, Puma now has five axes of
motion, which correspond to a human's waist rotation, shoulder rotation,
elbow rotation, wrist bend, and rotation. The new robot exhibits extraordi-
nary accuracy in its ability to repeat positions within 0.004 in. (0.1 mm).
Manipulator movements are accomplished by servomotors under the control
of an LSI-11 microcomputer. The robot is trained with a teach module, an
operator console, or a combination of both. Puma is a light- to medium-
payload robot with an end-effector capacity of 5 lb. Unimation, Inc. is
currently developing a machine vision system for Puma to enhance its versatility in manipulation of disoriented assembly-line objects.

Computer-controlled robot welding, once limited to automobile assembly welding, now moves into the shipyards with the aid of Unimation,
Inc.'s Apprentice robot shown in Figure 8.14. This is a portable robot that
can be set up on-site or in cramped quarters, such as ship interiors, to perform difficult vertical and out-of-position welds for extended periods. Once
the Apprentice robot has been programmed, it can be left alone to perform
all welds within its reach. The programmed welding path can be straight,
curved, or compound curved over a span of 13 ft. The Apprentice robot is
ideal where large quantities of weld metal must be deposited, as it can produce weaving or multipass welds of consistently high quality.

The Unimation, Inc.'s series 2000 industrial robot is one of the most
widely used robots today. Its industrial applications include spot welding,
investment castings, assembly-line operations, part transfer, and very hot
materials handling. Figure 8.15 shows the Unimation series 2000 robot performing an assembly-line function. This is not so unusual, but the series

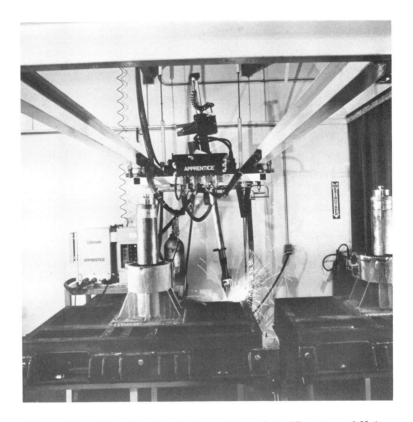

Figure 8.14   Unimation Inc.'s Apprentice robot. (Courtesy of Unimation, Inc.)

2000 robot is capable of moving conveyor-belt synchronization that enables it to perform its task with the workpiece in motion. The moving conveyor belt synchronizing system features include:

1. Coordinate translation based on known skewing of the conveyor belt or the workpiece on the conveyor
2. Injection of a subroutine, such as wiggling a spot welder gun, which has become stuck to the weld spot
3. Translation of robot motions into X, Y, or Z coordinates as a function of conveyor position for handling workpieces on the conveyor or extracting workpieces from deep fixtures

The Unimation, Inc.'s series 2000 robot has a load capacity of 450 lb at full arm extension. It has a 10-ft reach capacity and can span a 20-ft radius doing jobs involving massive workpieces or tools. Its repeatability accuracy is demonstrated in its ability to perform 1024 sequential steps to an accuracy

Figure 8.15 Unimation Inc.'s series 2000 heavy duty industrial robot. (Courtesy of Unimation, Inc.)

Figure 8.16 Unimation Inc.'s model 2000 robot showing long reach capability. (Courtesy of Unimation, Inc.)

Figure 8.17 Unimation Inc.'s series 2000 robots on an automobile welding assembly line. (Courtesy of Unimation, Inc.)

of 0.08 in. (2.0 mm). Figure 8.16 demonstrates the series 2000 robot's long-reach capabilities when used for cooling hot investment castings.

Industrial robots do a better job than human beings in assembly-line spot welding. The combination of moving conveyor belt synchronization and 1024-step sequence programming capability makes it an ideal assembly-line welding tool for the automobile industry. Figure 8.17 shows an "army" of Unimation, Inc.'s heavy-duty series 4000 industrial robots lining both sides of an assembly conveyor where automobile bodies move for spot welding. Each robot welder performs only X number of welds in the time an automobile body remains in the robot's work zone. The battery of welding robots can thus complete all specified spot welds at faster conveyor speeds than is possible with human welders.

CINCINNATI MILACRON INDUSTRIAL ROBOTS

According to the Robot Institute of America (RIA), by the mid-1980s there were more than 9000 industrial robots operating in manufacturing facilities throughout the United States and 100,000 in Japan. The growth in the U.S. robot market indicates an increase in industry's use of robots of 80,000 to 100,000 in the United States and 1,000,000 in Japan by the year 2000. These figures do not include computer-controlled metal, plastic, or wood-working machinery.

Figure 8.18 Cincinnati Milacron T3-586 industrial robot. (Courtesy of Cincinnati Milacron.)

Cincinnati Milacron is one of the major manufacturers of both metal-working machinery and industrial robots. Their computer-controlled series T3 industrial robot, shown in Figure 8.18, has become an industry standard for computer automation and has earned the reputation of being the workhorse of the metal fabrication industry.

The Cincinnati Milacron computer-controlled T3 industrial robot is produced in two models, the T3-566 standard and the T3-586 heavy duty. The T3-566 and T3-586 have unique joint-arm construction to enable the manipulator to have the flexibility to perform in difficult-to-reach places. Each of the manipulators six axes is driven by its own electrohydraulic servo system. Five of the axes use compact rotary actuators built into each joint and one axis is driven by a pivoted cylinder. This method of construction provides a backlash-free system capable of high torque, speed, flexibility, and accuracy over 240° of movement. Each axis has its own position feedback device, consisting of one resolver and tachometer to assure repeatability and precise manipulator positioning to plus or minus 0.050 in. (1.25 mm). Control of the action and positioning of each of the six axes is accomplished by the Cincinnati Milacron Acramatic microprocessor-based computer, which provides infinitely variable six-axis positioning and control path (straight-line) motion between programmed points. All of the T3 motions are referenced to the tool center point (TCP), a discrete point at a selectable distance from the manipulator where the tool meets the work. Figure 8.19 illustrates the T3 manipulator flexibility and the Acramatic control computer.

The Cincinnati Milacron T3 robots are taught their jobs by an easy-to-use Teach Box, as shown in Figure 8.20. The operator does not need prior computer experience or to perform mathematical calculations, just a knowledge of the physical job to be performed. The Teach Box enables the operator to program the T3 from the best vantage point. An offset branching function further simplifies the teaching in that a series of repetitive moves can be taught as a subroutine just once, then retained in memory for later recall when the same task comes up again. Those jobs the T3 performs that require lengthy teaching sessions or are reoccurring can be stored on magnetic data tape or disk for future use.

Comparison of T3 566 and 586 Specifications

Specification	T3-566	T3-586
Load capacity (lb)	100	225
Position accuracy (in.)	0.050	0.050
Maximum horizontal sweep (deg)	240	240
Maximum horizontal reach (in.)	97	102
Floor-to-ceiling reach (in.)	0–154	0–154
Maximum tool velocity (in./s)	50	35
Pitch (deg)	180	180
Roll (deg)	270	240
Yaw (deg)	180	180
Position points in memory	450 std.	700 std.

The Cincinnati Milacron T3 robots are widely employed in the automobile industry in computer-controlled spot welding of automobile bodies as they move past the robots on a conveyor. Figure 8.21 shows a T3-566 robot in the act of spot welding an automobile body as it moves along an assembly-line conveyor. Its spot welding and effector contains three welding tips to enable the robot to produce difficult-to-reach welds without having to change its welding end effector tool repeatedly. Figure 8.22 shows how the T3-586 robot reaches through an automobile's windshield and rear window areas to perform interior spot welding. The figure clearly shows the absence of human workers in the areas where industrial robots are on the job. Figure 8.23 shows the T3-586 robot's welding end effector in operation welding the base of a computer mainframe and the welding wire supply spool mounted at the elbow end of the manipulator.

The Cincinnati Milacron computer control robots can do more than welding jobs in industry. Figure 8.24 shows the T3-566 robot in the act of removing a part formed in a hydraulic press. Once removed, the part will be placed on a conveyor belt and sent on to its next fabrication step. In this application, the T3 robot serves not only as a parts-handing robot, but its use removes human workers from high-risk hazardous jobs such as hydraulic and

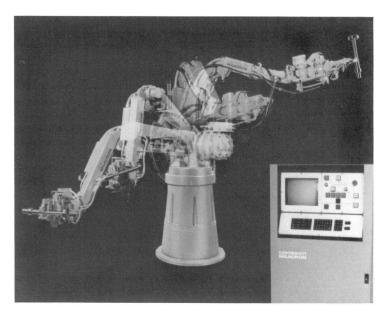

Figure 8.19 Flexibility of manipulator movements of a Cincinnati Milacron T3-566 robot and its Acramatic computer controller. (Courtesy of Cincinnati Milacron.)

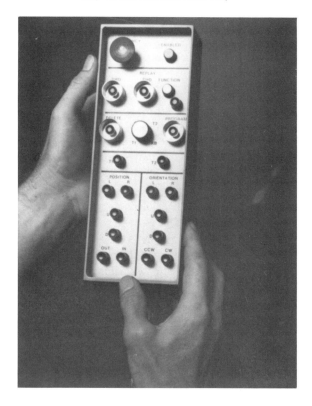

Figure 8.20 Cincinnati Milacron's Acramatic computer Teach Box. (Courtesy of Cincinnati Milacron.)

Figure 8.21 Three-headed spot welder end effector used by the T3-566 robot for automobile assembly line interior welding. (Courtesy of Cincinnati Milacron.)

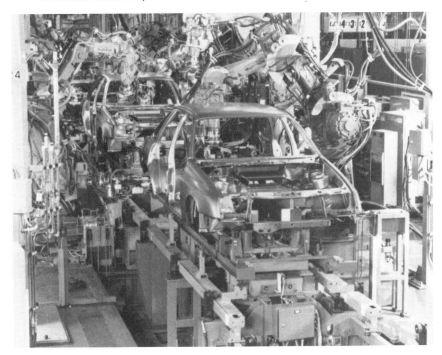

Figure 8.22 Assembly line welding with a Cincinnati Milacron T3-586 robot reaching through the windshield opening to perform interior welding. (Courtesy of Cincinnati Milacron.)

Figure 8.23 T3-586 shown welding a computer main frame and manipulator at 45-degree yaw axis position. (Courtesy of Cincinnati Milacron.)

Figure 8.24 T3-566 robot removing metal stamping parts from a hydraulic press. (Courtesy of Cincinnati Milacron.)

drop press forming operations, which cause many industrial accidents each year. Figure 8.25 graphically illustrates how welding robots are used to protect human workers. Electric arc welding produces harmful ultraviolet radiation to human eyes, and molten metal, ejected during the welding operation, produces painful burns on hands, arms, and anywhere they fall. Indeed, most human welders protect themselves on the job by wearing protective eye shields and spark-protective clothing and gloves—at least they are supposed to by company safety regulations. Still there are hundreds of industrial welding accidents every year—and that should tell us something.

Figure 8.26 shows how the computer-controlled T3-566 robot becomes a perfect companion to the Cincinnati Milacron Cinturn 12 universal NC turning centers. The robot is used for handling material and finished parts and the Cinturn to machine the parts. This combination results in a sophisticated, multimachine manufacturing unit capable of producing large quantities of precision workpieces continuously and with minimal human responsibilities.

Note that the T3's end effector has a two-part or dual-function gripper. One removes the completed machined part from between the Cinturn's center driver and tailstock, while the second gripper positions the new workpiece for machining. The T3 robot then places the completed machined part on another conveyor belt and picks up another workpiece from the incoming conveyor belt.

In any computer-controlled machine or robot application, the apparent

Figure 8.25   T3-566 robot showing how robots reduce h
conditions in arc welding. (Courtesy of Cincinnati Milacrc

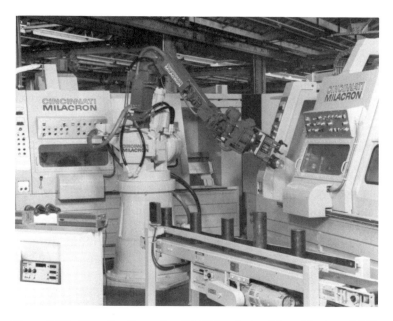

Figure 8.26 Coordination of a T3-566 robot with a Cincinnati Milacron computer controlled Cinturn 12 metal turning centers machine. (Courtesy of Cincinnati Milacron.)

results is high volume, very low rejection rate, and overall product improvement. What about the human workers that these machines replace? In a majority of cases, the displaced human jobs are those most human beings would prefer not to be assigned to in the first place. Frequently, the jobs are dirty, extremely hot, and definitely hazardous. These are the jobs that the industrial robots take over. What about the machinists when the computer takes over their milling machines or metal lathes? Generally, machinists are on piecework and their pay is based on the number of acceptable finished parts. In this case the computer-controlled metalworking machine drastically reduces the rejection rates, from 12.5 to 15% down to less than 1%, and simultaneously increases the total number of parts finished per shift. In this case the machinist actually does less work and receives more pay because of the production accomplished by the machine, technically under a machinist's control.

The major industries, where robots and computer-controlled machines have replaced human workers, provide in-house retraining programs or underwrite off-site retraining courses to upgrade the employee's ability to continue employment with the company. In this manner, machine- or robot-displaced employees are given an opportunity to retain their employment with better working conditions, take-home pay, and benefits.

CHAPTER REVIEW QUESTIONS

1. According to the dictionary definition of a robot, what are the two features included in the definition that are not part of an industrial robot?
2. What are the three basic components of an industrial robot?
3. Name the four basic designs of robot movements.
4. Name the six axes of robot movements.
5. Y-axis coordinate limits are given as $+120°$ and $-70°$. To what point are these coordinates referenced?
6. List three robot work functions that determine the selection of axes power choice.
7. What is the meaning of resilience?
8. Name two forms of pressure used to power robot axis movements.
9. What function will servomotion control have that enables the computer to know exactly how much movement or travel has been made?
10. Which robot would have the higher payload, assuming that each robot has the same constructural design and overall weight, one with three axial movements or one with six axial movements?
11. Hydraulic- and pneumatic-operated devices require what kind of plumbing?
12. Which would require greater space on a manipulator, an electric motor or a hydraulic rotary actuator?
13. How many axes of movement does the Unimation, Inc.'s Puma robot have?
14. The Apprentice welding robot can perform straight, curved, or compound curve welding over what length of workspace?
15. The Unimation series 2000 robot can perform how many sequential steps?
16. To what degree of accuracy can a series 2000 robot repeat each sequential steps?
17. If an industrial robot can perform a given number of spot welds in the period that an automobile body is within its reach, why would a large number of robots be used on either side of a welding conveyor and each robot welding less than it capability?
18. What does TCP mean in reference to a robot's end effector?
19. What must one know to teach a T3-566 robot to perform a task?
20. Which make of robot has the lowest repeatability tolerance, Unimation, Inc.'s or Cincinnati Milacron's?
21. Which Cincinnati Milacron robot has the larger payload: the T3-566 or the T3-586?
22. How do spot-welding robots get inside a car's body to perform welding operations?
23. How can human workers, replaced by robots, retain their jobs?

GLOSSARY

Actuator. A cylinderlike mechanical device that can transform hydraulic or pneumatic pressure into a linear or rotary motion.

Arc welder. A method by which metals are joined or fused together by heat produced by an electric arc generated by high current emitting from a fusible electrode to the junction point of the metals to be welded.

Axis. A movement centered around a given point or axis. In robotics, the Z axis refers to vertical movement, the X axis refers to base rotational movement, and the Y axis refers to the reach movement centered on the axis of the robot's fixed base.

Backlash. The error factor associated with the meshing of gears derived from the movement between the driven gear teeth and the driving gear teeth. Backlash tolerance is typically 0.002 in. per gear.

Configuration. The geometrical arrangement of components that gives an assembled device its overall shape and physical appearance.

Coordinates. The specific point, vertical, horizontal, and angular, that determine any given location. Navigation by ship or aircraft employs the coordinates of latitude and longitude to determine the location of the craft.

End effector. The portion of a robot manipulator that functions like the hand of a human arm.

Gripper. A movable section of an end effector which functions similar to the fingers on a human hand engaged in gripping or picking up an object.

Horsepower. A measurement of work associated with electrical motors or steam or internal combustion engines; related to 747.6 W electrical power and 500 ft-lb per second, where 1 hp = 550 lb could be lifted 1 ft in 1 s.

Hydraulic. The use of low-viscosity oil as a pressure-transfer medium and pressure amplification by differences between input pressure area and output pressure area within a cylinder-shaped device. A device operated by water or fluids under pressure.

Manipulator. A section of a robot corresponding to the arm section of a human being from the elbow to the hand.

Payload. The maximum load weight capacity of a mechanical moving device such as a robot, forklift, or dump truck.

Pitch. An up-down movement of a robot's end effector, similar to the up-down action of a human wrist.

Pneumatic. A source of power using pressure in the form of compressed air, gases, or a vacuum.

Resilience. The ability of a device to "give" instead of breaking when hard contact with a solid surface.

Robot. A machine or device controlled by a computer to perform work functions formerly performed manually or with manually controlled machines.

Roll. A particular circular movement of a robot's end effector similar to the rotating function of the human wrist; typically, 240° circular movement of the end effector.

Task. A specified movement or step to be accomplished in series with other tasks to accomplish or complete a given objective or job.

TCP. Tool center point: a discrete point at a selectable distance from the manipulator where the tool meets the work surface.

Yaw. A movement of a robot's end effector similar to a 180° left-right movement of a human wrist.

9

Fiber Optic
Transmitters-Receivers
and Data Links

Whatsoever man can envision, in time he can accomplish.
Unknown

For more than 4500 years all communications over distances beyond the range of the human voice depended on transmission methods we could either hear or see. Historically, the ancient Chinese used lanterns for signaling around 2000 B.C. and around 450 B.C. the Greeks developed the *heliograph* to send signals by reflecting the sun off a mirrorlike surface. We know that the American Indians developed expertise in the use of smoke signals; Africans, the sound of drums; and ships at sea, signal flags as a means of communicating. It was not until the late nineteenth century that Samuel Morse developed a system for transmitting code signals, which represented alphanumerical characters, by means of wires connecting two systems; and Alexander Graham Bell patented a method by which the human voice could be transmitted from one point to another by connecting wires: the telephone. Communication progress was not confined to the United States during this period, for across the Atlantic Ocean, a German physicist named Heinrich Hertz discovered the principles of electromagnetic radiation which became the vehicle for present means of communicating by radio and television. It is ironic that our means for transmission of data seems to be rediscovering the wheel, so to speak, by returning to the use of light for transmission of binary or analog data through fiber optic filaments instead of in electrical form transmitted over wire.

Radio as a means of communication and entertainment is little more than 60 years old, black-and-white television approaching 40 years, and color television barely over 30 years. We have learned from radio and television that weak or low-power signals produce poor reception in the form of

205

TABLE 9.1
Comparisons between Radio,
Television, and Fiber Optics

Bandwidth	
AM radio	10×10^3 Hz
FM radio	20×10^3 Hz
Color television	4×10^6 Hz
Fiber optics	2×10^7 Hz
	(2 megabytes)

Transmitter Power Level	
AM and FM radio	5-100×10^3 W
Television	1-3×10^6 W
Fiber optics	1-2×10^{-7} W

noise, static, loss of stereo, and low-level audio. In television, low signal levels result in loss of color, ghosts, and snow in the image. The threshold signal strength varies from 2 to 150 mV for radio reception and 1200 to 1500 mV for class A television reception. Compared to fiber optic signals, these threshold signal levels are enormously high, on the order of a thousand million times greater (2×10^{-12} to 2×10^{-3}). Bandwidths, on the other hand, are a mere fraction of that of fiber optics, as shown in Table 9.1. Table 9.1 also compares transmitter power levels. On this basis of comparison, AM and FM radio transmitter power runs 10 million times higher than fiber optic transmitters and television transmitter power 10 billion times higher.

DATA TRANSMISSION AND RECEPTION VIA OPTICAL FREQUENCIES

In Chapter 2 we pointed out the advantages of fiber optic data links over shielded wire in areas where electrical and radio-frequency interference levels are at significantly high levels which could induce data distortion or loss. Mentioned, but not emphasized, was the magnitude of data transmission through fiber optic cables in comparison to shielded cables; on the order of 1000 times more data can be transmitted through fiber optic cables. Transmission attenuation again favors the fiber optic cables over shielded cables; 84% optical power loss (8 dB) per km to 5×10^{-13} power loss (125 dB) per km for certain types of shielded cables. The only obvious advantage held by shielded cables is the ability to operate bidirectionally through alternating transmit and receive modes.

The basic structure of a fiber optic data transmitter and receiver is illustrated in Figure 9.1. An analog or digital electrical signal is fed to the source driver amplifier/modulator, which causes the current to vary at the data rate as it passes through the optical source. This produces an increase in the intensity of the optical source, which also varies at the data rate. This effect is

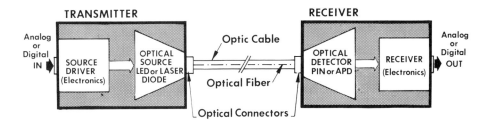

BASIC FIBER OPTIC DATA LINK

Figure 9.1   Basic fiber optic transmitter and receiver diagram. (Courtesy of American Photonics.)

called *intensity modulation*. The intensity-modulated light is then injected into the optical fiber and transmitted to an optical receiver. Upon arriving at the optical detector, either a PIN diode or an avalanche photo diode (APD), the variations in the received intensity are converted back to analog or digital electrical signals.

FIBER OPTIC TRANSCEIVERS AND TRANSMITTERS

Figure 9.2 illustrates the American Photonics, Inc.'s TR1000 fiber optic transceiver. It combines an optical data transmitter and receiver into a small dual-in-line package (DIP) which is not unlike the familiar DIP logic gate package except for size. It measures 0.75 in. high, 1.8 in. wide, and 2.8 in. long. There are fourteen connecting pins arranged in two rows of seven pins on 0.4-in centers.

The transmitter output and receiver input are short fiber optic "pigtail" stubs for connection to a dual-fiber optic cable. The TR1000 transceiver uses an intensity-modulated LED as the optical source and a PIN photo diode as the receiver detector. These devices produce a transmitter bandwidth of 24 MHz (measured 6 dB down) and a receiver bandwidth of 27 MHz (measured 6 dB down).

TR1000 Transmitter Operation

The schematic drawing in Figure 9.2 shows a data differential amplifier, formed by transistors Q1, Q2, Q3, and Q4, in an active biasing configuration. Active biasing enables the transmitting LED to operate over a wide range of power supply voltages (V_{cc} and V_{ee}) and current levels. Modulation levels are set by adjusting the voltage at pin 5 to the desired current level through the LED. Transistors Q3 and Q4 are closely matched so that under quiescent conditions, the emitter currents of Q1 and Q2 are the same. A

TR1000 Fiber Optic Data Link Transceiver

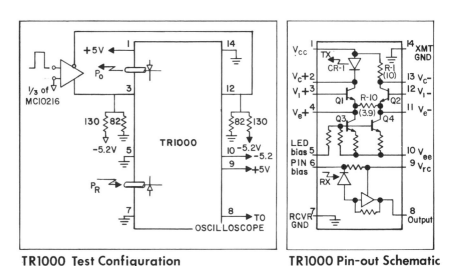

TR1000 Test Configuration   **TR1000 Pin-out Schematic**

Figure 9.2 American Photonics TR1000 fiber optic transceiver. (Courtesy of American Photonics.)

temperature-compensating network is used to bias Q3 and Q4 to offset ambient-temperature effects on the biasing of transistors Q1 and Q2. Temperature compensation does not compensate for variations in output LED (CR-1) optical power, which varies approximately 0.3% per degree Celsius.

Current through LED CR-1 is determined by measuring the voltage drop across resistor R1 instead of taking a direct measurement across CR-1 because the normal 1.8-V drop across the LED is not linear with its forward current and will vary with a data signal input to transistor Q1's base. To prevent Q1 from becoming forward biased, the peak input voltage at pin 3 should never be more than $V_{cc} - 3$ V.

Transistors Q3 and Q4 bases must not be forward biased by applying a high voltage on pin 5. Should Q3 and Q4 bases be forward biased, it would cause the LED to draw excessive current and be destroyed.

The LED's optical power will be proportional to the collector current of Q1 and the level of current is set by the amount of voltage on pin 5. The logic or analog current level is determined by the differential input voltage divided by R10. The absolute maximum LED forward current is 100 mA at 100% modulation. This requires a maximum input differentiated level of 780 mV. Where input peak-to-peak voltages are greater than the 780-mV differential level, attenuation of the input voltages must be performed.

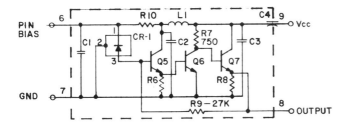

Figure 9.3   TR1000 fiber optic
receiver schematic drawing.
(Courtesy of American Photonics.)

TR1000 Receiver Operation

In Figure 9.2, the receiver section was simplified to conserve schematic space and the trans-impedance amplifier was shown only as an amplifier symbol. The full receiver schematic is shown in Figure 9.3 and corresponds to the receiver operation description.

The receiver is a three-stage amplifier with a feedback resistor. Upon receiving an input signal, the PIN (CR-3) internal resistance varies in step with the amplitude of the analog or digital data signal. This produces a voltage variation in the forward bias on the base of Q5. Transistor Q5 is connected in a class A amplifier in emitter-follower configuration and will develop a voltage across resistor R6. When a data signal is received by the PIN, the emitter voltage of transistor Q5 will vary in phase with the input data signal. The no-signal voltage across emitter resistor R6 provides the forward class A bias to transistor Q6, and its voltage variations constitute an in-phase amplification of the input data signal. Transistor Q6 is a class A amplifier in a grounded emitter configuration. The data signal variations in its forward bias voltage produce identical variations in the collector current of Q7. This results in a corresponding voltage-drop variation across resistor R7 to produce and out-of-phase amplified image of the data input signal. The collector voltage of Q6 is direct coupled to the base of transistor Q7 to simultaneously provide class A forward bias and an amplified data signal which is taken from the emitter of Q7 as the amplified output signal. Resistor R9 is connected to the emitter of transistor Q7 and returns a small portion of the output signal to the base of transistor Q5 as an inverse feedback voltage. The feedback voltage variations are 180° out of phase with the input signal phase and serve to maintain uniform amplification and reduce distortion by lowering the input transistor's (Q5) gain factor. The three transistor trans-impedance amplifier circuit's quiescent voltage is V_{be} Q5 + V_{be} Q6 or the base-to-emitter voltage of Q5 and Q6, which typically runs 1.4 V.

The simplest digital receiver would be where the TR1000 receiver is coupled to an ac comparator. In this configuration, the duty cycle of the receiver must not vary greatly from a 50% duty cycle. The received optical power must be sufficient to produce an output voltage at pin 8 that exceeds the threshold of the comparator. For a comparator whose threshold is approximately 5 mV, the average optical input power to the receiver will require a minimum of 1 μW. At this power level, the signal-to-noise ratio is about 27 dB, which is approximately 11 dB above the theoretically required sensitivity for a 10^{-9} bit error rate. This circuit allows a minimum system margin of

$$\text{system margin} = 10 \log \frac{P_o}{P_r}$$

where P_o = average output power
P_r = average received power

Therefore,

$$\text{system margin} = 10 \log \frac{40 \ \mu\text{W}}{1 \ \mu\text{W}} = 26 \text{ dB}$$

Figure 9.4 shows the American Photonics TR1000 transceiver, which has a transmitter optical output power of approximately 40 μW output when the LED is driven by a 100-mA forward current. The 26-dB system margin makes it capable of transmitting over 3 km, provided that the fiber optic link has an attenuation of no more than 6 to 8 dB/km. With transmission links longer than 3 km or attenuation greater than 8 dB/km, a linear voltage amplifier must be used between the TR1000 and the comparator in order to increase the output voltage greater than the threshold of the comparator.

When low-level signals are to be detected, overall performance in signal sensitivity can be improved by the use of a comparator whose threshold is signal dependent or the receiver employs an automatic gain circuit (AGC). This would ensure that the signal level is maintained above the threshold of the comparator and that the noise level is as far from threshold as possible. Signal sensitivity can be further increased if it passes through a low-pass filter before going to the comparator.

Figure 9.4 Photograph of the TR1000 transceiver. (Courtesy of American Photonics.)

INDIVIDUAL TRANSMITTER AND RECEIVER SYSTEMS

The majority of computer control data links run considerably less than 1 km, more on the order of 30 to 50 m. The low attenuation of optical cables over this distance will allow the data transmitters to function at lower optical power while maintaining the same system margin. In most applications, there can be a significant system margin improvement with the use of lower optical power transmitters when the data links are under 1 km in length.

It is easy to confuse the relationship between transmitting electrical signals over shielded wire cables and transmitting optical signals over fiber optic cables. With shielded wire cables we are concerned with transmitter power, cable attentuation, frequency response, and receiver sensitivity. Wire diameter has no appreciable effect on impedance-matched frequencies and no effect on receiver sensitivity except the signal attenuation over given distances.

The diameter of the optical fiber is directly related to the minimum optical power requirements, the speed of data transmission (baud) or signal frequency bandwidth and receiver sensitivity requirements. Table 9.2 illustrates this relation for a typical transmitter/receiver combination for three optical fiber diameters.

Table 9.2 is based on separate fiber optic transmitters and receivers produced by the Burr-Brown Corporation. The first column, headed "T/R Combo 1," combines a model 3713T with a model 3712R receiver for maximum low-speed performance with data links longer than 1 km. The high power output of the 3713T transmitter eliminates the need for signal repeaters when used with the model 3712R receiver's superior sensitivity. The second column, headed "T/R Combo 2," shows the low-speed performance of model 3712T transmitters and model 3712R receivers when used with data links under 1 km. The third column, headed "T/R Combo 3," shows the high-speed performance of model 3713T transmitters and model 3712R receivers.

TABLE 9.2   Optical Fiber Diameter Power Comparisons

Fiber Diameter (μm)	T/R Combo 1		T/R Combo 2		T/R Combo 3	
	Minimum	Typical	Minimum	Typical	Minimum	Typical
	Transmitter Output Power					
1016	10	15	2.5	3.5	10	15
368	1.5	3.5	0.33	0.46	1.5	3.5
200	0.7	1.5	0.10	0.7	0.7	1.5
	Receiver Sensitivity (nW)					
	5	2	5	2	30	15
	Data Link Bit Rate (baud)					
	0–20k	0–25k	0–20k	0–25k	0–200k	0–250k

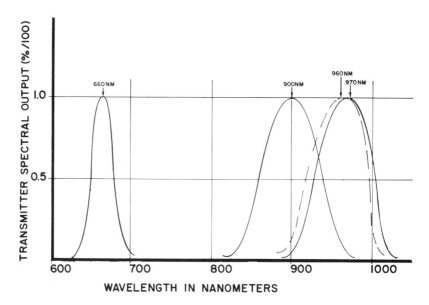

Figure 9.5 Fiber optic transmitter LED spectral output graph. (Courtesy of Burr-Brown Corp.)

When comparing shielded-wire cable to optical fiber cable, it is important to note that the smaller-diameter optical fibers require lower optical transmitter power to transmit data over equal-length data paths. The selection of the transmitter's optical frequency has an important bearing on the signal attenuation of the optical fiber; the rated attenuation per kilometer is based on 660, 900, 960, and 970 nm. Graded-index optical fibers have the lowest attenuation when the transmitter operates at IR frequencies of 1200 to 1400 nm. Typical transmitter spectral power is shown in Figure 9.5 for LED optical sources. Note that the bandwidth is quite narrow at the 660-nm frequency. However, it is wide enough for transmission of low-speed data. The optical source between 900 and 970 nm has a bandwidth sufficiently wide for high-speed data transmission in excess of 250K baud.

The Burr-Brown fiber optic transmitter schematic, shown in Figure 9.6, is the same basic configuration for the model 3712T and model 3713T transmitters. They differ only in output optical frequency, output optical power, and rise/fall time. Each transmitter uses a Schmitt trigger exclusive-OR gate (G-1) for noise immunity. Its logic is configured so that phasing of the transmitter is pin-programmable. When the transmitter is connected to phase 0, the light output is in phase with the digital input signal. The LED source is on when the digital logic is high. By connecting the transmitter to the phase 180 pin, the LED source is on when the logic input is low. The advantage of the phase 180 connection is that the LED would be ON when the data input is idle and would simplify locating breaks in the optical cable. It is also par-

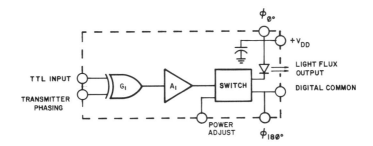

Figure 9.6 Burr-Brown FOT 3712T and 3713T fiber optic transmitters. (Courtesy of Burr-Brown Corp.)

ticularly useful in the transmission of asynchronous data with an idle state of a TTL low. Both model transmitters operate at the same 5 V dc source voltage and draw currents of 40 to 60 mA. The differences between the transmitters are listed in Table 9.3.

Higher-power fiber optic transmitters normally operate most effectively in the near-infrared portion of the spectrum. Figure 9.7 shows the simplified schematic of the Burr-Brown FOT-110KG-IR transmitter, which has a minimum optical power of 50 μW, typical output at 90 μW at 880 to 910 nm, and data transmission at 0 to 0.75 megabyte.

The FOT-110KG-IR transmitter consists of digital input gates, a current mirror, and an LED. The digital gates are low-power Schottky TTL open-collector exclusive-OR gates. In operation, when the enable is high ("1"), the data input will control the state of the LED. The data-LED phase relationship is controlled by a logic 1 or 0 to the phase control, as shown in Table 9.4.

Biphase modulated data can be generated by applying a clock signal to the phase control input and clock-synchronized data-to-data input. This technique reduces the baud rate but is useful in generating an ac signal.

The LED is driven by a ratioed current mirror as shown in Figure 9.7. When the voltage at the base of Q1 is high, the LED is ON. If the R-external = 0, the LED will be at a maximum current of 108 mA. Increasing the value of R-external will reduce the LED's current and output power and extend the life of the LED. In addition, the LED can be intensity modulated directly by the voltage level applied to the amplitude-modulation (AM) input. The AM signal can be audio or low-frequency RF up to 1 MHz. An input

TABLE 9.3 Burr-Brown Optical Transmitter Variations

Variation	Model 3712	Model 3713
Optical frequency (nm)	670	660
Maximum output power (μW)	3.6	15
Rise time (ns)	120	40
Fall time (ns)	50	75
Bit rate (baud)	0–2M	0–2M

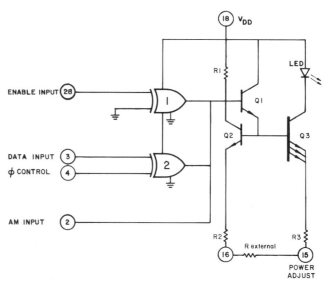

FOT-IIOKG-IR TRANSMITTER

Figure 9.7 Burr-Brown FOT-110KG-IR fiber optic transmitter schematic. (Courtesy of Burr-Brown Corp.)

voltage level of 1.1 V peak to peak at the AM input will produce intensity modulation of the LED at 100% modulation. Figure 9.8 shows the Burr-Brown FOT-110KG-IR transmitter and the FOR-110KG receiver.

Fiber Optic Receivers

Figure 9.9 is an abbreviated schematic of the Burr-Brown Corporation's 3712R fiber optic receiver. It differs from the TR1000 receiver in that it contains a current-to-voltage converter followed by a voltage amplifier and a comparator to provide separate analog or TTL outputs and incorporates an automatic threshold circuit. In operation, the input optical signal is fed to a PIN photodetector which is connected in a photovoltic mode for maximum sensitivity. A low-bias-current FET input current-to-voltage converter transforms the PIN current into a voltage (V_a). This voltage is further amplified by A_2 and transferred to a comparator A_3 as V_b. Here it is compared to the

TABLE 9.4   Transmitter Truth Table

Enable input	Data input	Phase input	LED
0	X	X	Off
1	0	0	Off
1	1	0	On
1	0	1	On
1	1	1	On

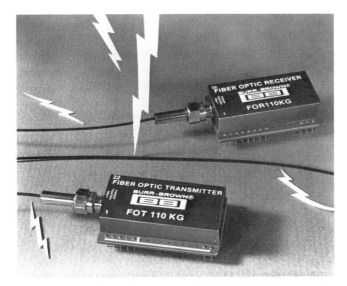

Figure 9.8 Photo of the Burr-Brown FOT-110KG-IR transmitter and receiver. (Courtesy of Burr-Brown Corp.)

threshold voltage V_t. For maximum noise immunity, the threshold voltage should be set to a level midway between the high and low levels of input regardless of the light-level input to the PIN detector. The automatic peak de-

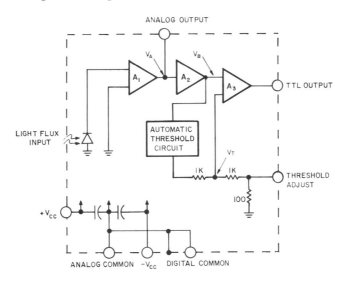

Burr-Brown 3712 Receiver

Figure 9.9 Schematic of the 3712R receiver. (Courtesy of Burr-Brown Corp.)

tector function of the automatic threshold circuit receives a voltage pulse (V_b) which is stored in the automatic threshold circuit, divided in half, and supplied to the comparator as the threshold input V_t. In this manner, V_t is a voltage corresponding to the midpoint of the light and no-light conditions at the PIN detector.

The automatic threshold circuit uses a capacitive hold technique where the threshold voltage (V_t) decays when light is removed from the detector. A no-light condition of approximately $\frac{1}{2}$ (a 1-baud data rate) can be used with no significant effect on noise immunity. If the input light to the PIN detector is cut off indefinitely, the voltage V_t will fall to the noise state, causing the TTL output to be subject to normal noise outputs. Normal threshold operation resumes on the first input light pulse. However, the initial transition at the TTL output may be uncertain for this first pulse, but after the first pulse activates the automatic threshold circuit there will be no uncertainty in the TTL output.

The schematic in Figure 9.9 is identical to both Burr-Brown fiber optic receivers, models 3712T and 3713T. They differ mainly in baud rates, sensitivity, and optical frequencies. Table 9.5 compares these characteristics.

The companion receiver to the FOT-110KG-IR Burr-Brown transmitter is the FOR-110KG receiver shown in schematic in Figure 9.10. The receiver has a bandwidth of dc to 1 MHz and is capable of receiving from 0 to 2 megabaud at an input signal level of 15 nW (-47-dB level). The receiver consist of a PIN photodetector and a trans-impedance amplifier followed by two comparators, one for data and one for squelch. The PIN photodetector converts the incident signal light from the optical fiber to a current corresponding to the optical power received. The signal current is then converted to a voltage by a low-noise high-speed trans-impedance amplifier. The amplifier's output voltage is then reconstructed to a TTL output by comparator 1. If the transmitter is amplitude modulated, an AM signal can be obtained at the trans-impedance amplifier's output.

The reference of the comparators is automatically set by the auto threshold peak detector. The threshold voltage at the link monitor output is equal to one-half of the amplifier's peak output level. This assures that comparator 1 will always switch with maximum noise immunity. Unlike fixed threshold receivers, this circuit maintains symmetry over a wide dynamic range. In addition, it provides an improved bit error rate with larger input signals. Since the FOR-110 is dc coupled, it allows far greater noise immunity than edge-triggered systems. The squelch function allows the data output to be

TABLE 9.5   Burr-Brown Receiver Comparisons

Function	Model 3712T		Model 3713T	
Data rate (baud)	0–20k	0–25k	0–200k	0–250k
Sensitivity (nW)	5	2	30	15
Spectral frequency (nm)	670		660	

FOR-110 RECEIVER

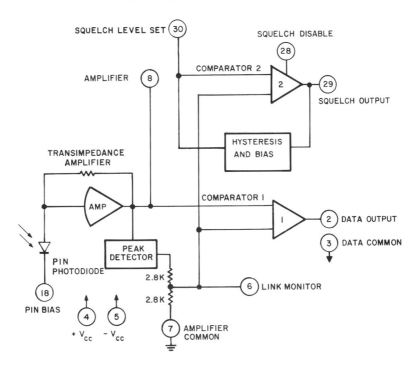

Figure 9.10   Schematic of the FOR-110KG receiver. (Courtesy of Burr-Brown Corp.)

cut off by the comparators when the optical input power falls below a preset level. This is accomplished by strapping the squelch out to the data out. With no external connections to the squelch level set, the squelch level will be approximately 20 nW. The squelch can be turned off by the application of a logic 1 to squelch disable. The link monitor output can be used to monitor the continuity of the link and to monitor the received data signals on an oscilloscope.

COMPUTER INTERFACING TO FIBER OPTIC TRANSMITTERS AND RECEIVERS

To interface a computer to fiber optic data links, a serial RS-232 or RS-422 I/O port is required for bidirectional transmission of analog or digital data between the computer and external optical transmitters and receivers. Connections between the interface and the fiber optic transmitter or receiver can

Figure 9.11   Belden Bit Driver multiplexer. (Courtesy of Belden Fiber Optics.)

be either electrical or optical via the appropriate dual-conductor shielded wire cable or dual-fiber optic data link cable.

Each of the fiber optic transmitters or receivers presented in this chapter may be operated by optical or electrical data signals. This feature enables

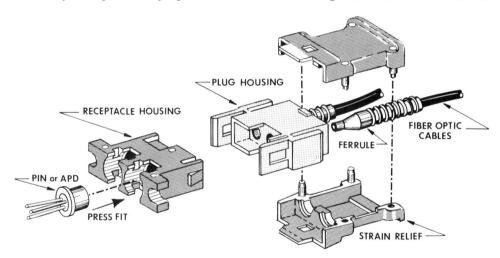

Amphenol Optimate Duplex Connector

Figure 9.12   Amphenol Optimate connector drawing. (Courtesy of Belden Fiber Optics.)

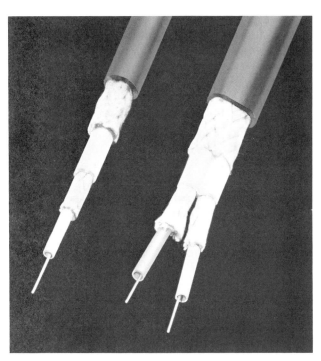

Figure 9.13   Belden heavy duty single and dual fiber optic cables. (Courtesy of Belden Fiber Optics.)

virtually any serial RS-232 computer I/O to interface the computer to a fiber optic system. One recent innovation in interfacing the computer to fiber optic data systems is the Belden Bit-Driver asynchronous and synchronous modems, which are single RS-232-C I/O ports containing an on-board optical transmitter and receiver to interface directly into duplex fiber optic data links. With the Belden Bit Drivers, duplex transmission can be transmitted up to 20,000 ft at operating speeds from 110 bits per second (bps) to 56 k bps. Figure 9.11 shows the Belden Bit Driver multiplexer, model 222006, which interfaces to eight different channels over a maximum distance of 6500 ft. The transmitter optical power (into a 300 micron core fiber) is rated at 400 μW at an optical frequency of 820 nm and the receiver sensitivity is rated at 1 μW at less than 10^{-9} BER (bit error rate). Connections to the Bit

TABLE 9.6
Operating Distances for Fiber Optic Cables (ft)

Belden number	Multiplexer	Modem
220002	—	3100
221002	—	3750
226002	6500[a]	5575
227002	3250	2540
227202	6500	5080

[a]Minimum distance is 1625 ft.

Driver modems or multiplexer are by Amphenol Optimate duplex connectors, as shown in Figure 9.12. The Optimate connector can be used with
300 micron fiber core filaments either in single-core or duplex-core cables,
similar to the Belden heavy-duty optical cables shown in Figure 9.13.
Table 9.6 shows the operating distances over which data can be transmitted
or received with the Bit Driver modems or multiplexers. A complete listing
of fiber optic data link equipment is included in Appendix F.

CHAPTER REVIEW QUESTIONS

1. What was the name of the device first used to send signals by light?
2. Who was the discoverer of electromagnetic radiation?
3. Which method of transmitting information has the widest bandwidth: AM radio, television, FM radio, or fiber optics?
4. What is the transmitting power range of AM-FM radio to fiber optic transmitters?
5. What are the typical power losses in decibels between shield cable and fiber optics
   per kilometer?
6. Define intensity modulation in respect to optical signals.
7. How does a TR1000 transceiver package compare to a DIP integrated-circuit package?
8. To what must the fiber optic data line be spliced to run a TR1000 transceiver?
9. Name two types of receiver photodetectors.
10. What is used for the transmitting optical source?
11. How many amplifier stages will a trans-impedance amplifier have?
12. How is a fiber optic systems margin determined?
13. What is the typical optical output power rating for fiber optic transmitters?
14. What is the normal receiver sensitivity for 10^{-9} bit error rate?
15. What advantage could be gained by the use of a separate transmitter and receiver over
    a single-package transceiver?
16. Does the diameter of the core optic filament have any effect on transmitter output
    power?
17. Is there an advantage in using high-output optical power when the data link is less
    than 3000 ft?
18. Which optical frequency would be best for 250-kilobaud data transmission: 660, 670,
    or 880 nm?
19. Which of the following dc supply voltages is used by the Burr-Brown fiber optic trans-
    mitters: +15 and −15 V dc, +20 and −17.5 V dc, +5 and −5 V dc, or +5 V dc and
    ground?
20. List the three sections of the FOT-110KG-IR transmitter.
21. The FOT-110KG-IR transmitter has a logic 0 on the enable input and nothing on the
    data or phase input. What is the state of the LED: ON or OFF?

GLOSSARY

AGC. Automatic gain control.

Alphanumeric. Characters and digital numbers used in writing the English language.

APD. Avalanche photo diode.

Attenuation. Reducing or lowering a signal level.

Baud. A term denoting the speed or rate by which data is transmitted or stored.

BER. Bit error rate.

C. Schematic symbol denoting capacitance.

Class A amplifier. An amplifier conducting throughout the 360° of each alternating frequency being amplified; a linear signal cycle; a linear signal amplifier.

CR. Schematic symbol denoting a semiconductor diode.

dB. Decibel; a logarithmic ratio between voltage current or power levels.

FET. Field-effect transistor.

Heliograph. A device by which signals may be sent as interrupted flashes of light by reflecting light from the sun with a movable mirror.

Hz. Hertz, a term honoring Heinrich Hertz for his discovery of electromagnetic waves; denoting frequency or number of alternations (cycles) of an alternating frequency per second.

IR. Infrared: the invisible optical energy below the visible red portion of the light spectrum; heat energy.

k. Abbreviation for the metric term kilo, denoting 1000.

km. Kilometer; 1000 m.

m. Milli, a metric term denoting 0.001 or 1×10^{-3}.

mA. Milliampere: denoting 0.001 A of an electrical current.

mSec. Millisecond; 0.001 s.

M. Mega, a metric term denoting 1,000,000 or 1×10^{6}.

MHz. Megahertz: denoting 1,000,000 cycles.

Modulation. A term indicating altering a dc voltage by superimposing an analog or digital signal onto the dc level, causing it to increase or decrease in step with the signal.

n. Nano: a term used to indicate the sum of 0.000000000001 or 1×10^{-12}.

nm. Nanometer; 0.000000000001 m.

ns. Nanosecond; 0.000000000001 s.

Pigtail. A short stub or section extending from an internal device circuit as a connection to an external device or circuit.

PIN. A semiconductor photodiode containing an intrinsic barrier between the P and N semiconductor junctions.

P_o. Power output.

P_r. Power received.

Q. A schematic symbol denoting a semiconductor device such as a transmitter.

Quiescent. A state during which no signal is being transmitted or received by an optical cable or device.

R. A schematic symbol denoting a resistor.

Rec. Receiver.

V_{cc}. Principal source voltage.

V_{dd}. A secondary source voltage lower in level or polarity than the V_{cc}.

V_{ee}. A secondary source voltage lower in level than V_{cc}.

V_{ss}. Normally associated with negative or ground-return connections of logic circuits.

10

Model Robots as Teaching Aids for Electromechanical Engineering

Little boys never grow up; their toys simply get more expensive.

There is a common misconception among educators that one can learn virtually anything by reading. Although this is true to a certain extent, there are exceptions. As an example, one could read everything about the piano and learn how it is made, tuned, and the musical notes for each key. However, without the skills developed through constant practice and the ability to read sheet music, it would be impossible to play a piano to the level of a concert pianist with the knowledge learned through reading. It is inconceivable that anyone would dare fly in an airplane if the pilot's sole training consisted of reading every book on airplanes and flying.

The use of model robots offers supplemental instructions to the mechanics of motors, gear trains, power, leverage, and force. At the same time, they provide hands-on experience in the electronics of external sensors and control methods. Robotic training aids are available for all levels of education from public schools through college. It is not surprising that the production of robotic training aids now exceeds that of commercial industrial robots.

One of the first robot training aids was the Turtle robot shown in Figure 10.1. It was designed for computer control via a 10-conductor umbilical cable: eight conductors connecting to the computers parallel I/O port and two to furnish power to the Turtle's electronic circuits and motors. It could be computer programmed to seek out a clear path through a complex maze. Its clear plastic dome served as an obstacle contact sensor, and each blind path encountered a contact sensor signal that would be relayed back to the

Figure 10.1   Original "Turtle" (TM) robot produced by Terrapin, Inc.

computer to be stored as an error path. Distance was calibrated by the number of revolutions of each wheel and also stored in the computer's memory. In this manner the Turtle could be run into the maze to locate and remember each turn error on its first try. Then, having thus located the maze's error paths, on its second try the Turtle could complete its passage through the maze error-free. Once the Turtle had mastered the maze, it could be placed on a sheet of paper and, with its on-board solenoid-activated ball-point pen, draw a miniature version of its path through the maze. Since the Turtle was available only as a kit, it also provided students with firsthand experience in the assembly of electronic components onto a printed circuit board.

Most training-aid robots serve multiple purposes of demonstrating motor-driven devices, control of motors, and control methods. A number of simple model robots, available through H&R Corporation in Philadelphia, Pennsylvania, demonstrate electronic motor control, gear trains, and control by infrared radiation or sound.

Figure 10.2 shows a Line Tracer robot, which has the capability of following a black line drawn on the floor or a large sheet of paper. The method of control is an IR LED transmitter and two IR receiver photodiodes. If the IR transmitter is centered on the black line, both wheels are driven forward. If the IR transmitter is off the black line, only one wheel receives forward driving power and the robot circles around the black line until it agains centers with the line. The assembly of the kit provides the student with hands-on experience in the assembly of motor-driven gear trains and small-component assembly with 2-56 machine screws and nuts.

Figure 10.3 shows the Piper Mouse, a sound-activated robot. The sound from a whistle causes the robot to move forward until it receives a second

Figure 10.2   "Line Tracer" robot controlled by infrared radiation.

whistle signal which causes the robot to stop. A third whistle signal causes the robot to turn clockwise; a fourth, counterclockwise; and it stops on the fifth signal. Like the line tracer robot, the whistle-controlled robot provides

Figure 10.3   Sound activated and controlled robot, "Piper Mouse."

the student with hands-on experience in small assembly and motor-driven gear trains.

Figure 10.4 is called the Space Invader robot. Its six moving legs enable it to climb over obstacles that would stop the two wheel-driven robots shown in Figures 10.2 and 10.3. It runs continuously once the power switch is switched ON. Should it encounter an obstacle, its IR sensor causes the drive motors to turn the robot continuously until the IR sensor indicates a clear forward path.

Figure 10.5 shows a programmable 4k RAM robot, Memocon, that is programmed via a seven-function teach pendant or via a parallel I/O port of a microcomputer. All four robots in this series are approximately 5 in. round (or square) and none over 3.5 in. high. Movement speeds are low enough to conduct laboratory experiments on the average-size lab bench.

Another step up the ladder of teaching-aid robot evolution, we find the computer-controlled robot RBX5 shown in Figure 10.6. RBX5's microprocessor is the National Semiconductor INS8073 with 8k RAM expandable to 16k. Its program language is Tiny BASIC Robot Control Language. Available software includes Alpha and Beta programs that enable RBX5 to learn from its experience. Optional software is packaged as plug-in modules which are plugged into a panel on RBX's back. RBX5 measures 13 in. in diameter and 24 in. tall, which provides ample interior room for additional user electronics,

Figure 10.4 "Space Invader" infrared obstacle detector robot.

Figure 10.5 Programmable 4k RAM robot, "Memocon," controllable by a teach pendant or computer. (Courtesy of Stock Model Parts.)

interfaces, and control functions. Interfacing to a computer is via an RS-232 serial I/O. RBX5's basic external sensors include eight tactile sensors extending around the middle of its body, an LED edge-detecting sensor located under its skirt at the front caster, and sonar for obstacle detection up to 35 ft. Two sealed lead-acid 6-V batteries provide separate power sources of 5-ampere-hour (Ar) for on-board electronics and 10 Ah for wheel-drive motors.

While the teaching-aid robots presented are fully capable of demonstrating a limited number of electromechanical principles, they fall short in

Figure 10.6 RBX5 robot showing how to install the plug-in program modules for its control functions. (Courtesy of RB Robot Corporation.)

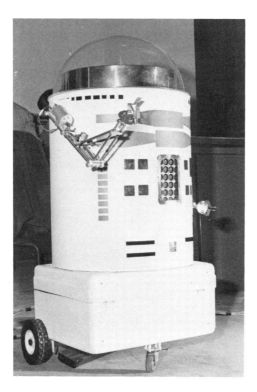

Figure 10.7 Original switch-box controlled Unicorn-One robot.

providing sufficient technical literature and tie-in demonstrations for college courses. A better approach would be to design a totally different type of robot as a means not only of teaching electromechanical principles but to provide ongoing projects that would offer the students hands-on experience in robot design and fabrication for a period of years.

Perhaps the most recognizable robot in this category, second only to R2D2 of Star Wars, is Unicorn-One, an acronym for "universal controllable robot." Unicorn-One was the subject of the longest-running construction article series ever published in *Radio Electronics* magazine (August 1980 through June 1981).

Figure 10.7 shows the original Unicorn-One, which was switch controlled for 10 functions: forward, reverse, shoulder actions, elbow actions, manipulator end effector actions, and body rotation. Turning actions were accomplished by operating a single wheel-drive motor; right wheel for left, left wheel for right. The tenth function was a relay that turned on a cassette recorder to play a recorded tape that gave Unicorn-One its voice.

Unicorn-One was a continuing student project from 1978 through 1984 at the Union County Career Center near Monroe, North Carolina. It underwent conversions from switch control to radio control, its voice changed from recorded cassette tape to student-operated wireless FM microphone,

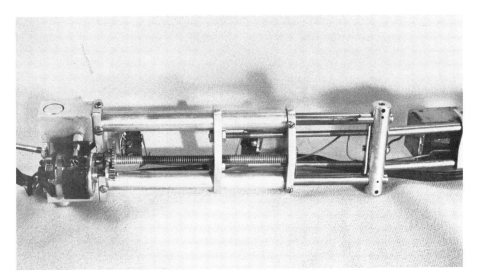

Figure 10.8 Telescoping manipulator used on the radio controlled prototype model of Unicorn-One.

and any function subject to mechanical problems was modified to a trouble-free operational state. The ultimate goal for Unicorn-One is to increase its number of action functions and to modify it to 100% computer control with speech synthesis and voice recognition. Unicorn-One's success is measured in the number of schools throughout the United States that have created their own class robot project and the additional skills mastered by the students in fabricating the metal manipulators, as illustrated in Figure 10.8. The only tools used to fabricate this manipulator were a hacksaw, electric hand drill, metal files, taps, and drill bits. The manipulator shown in Figure 10.8 has a telescoping action instead of the elbow action and its end effector can be rotated 360° in clockwise or counterclockwise directions.

DESIGNING A TEACHING-AID ROBOT

Student-project robots do not have to follow the designs of motion picture or television robots. In fact, they can range from dragon heads, as shown in Figure 10.9, to a computerized Apple Computer logo. Whatever the design, it should incorporate as many mechanical engineering principles as possible and the electronics for control or interface to a computer.

The three basic sections of any robot are the wheel support base, the superstructure, and the manipulator. Each section must be designed within the weight limits of the driving motors and the space required for on-board power and controlling electronic circuits. Internal accessibility to motors and

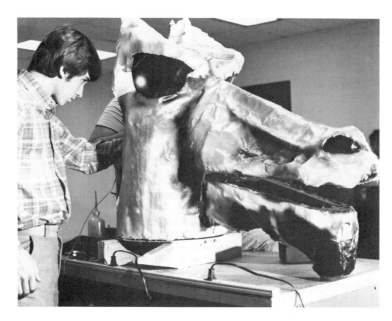

Figure 10.9   Union County Career Center student, Jack Plyler, adjusts his radio control dragon head robot.

electronic circuits is as important as structure integrity and strength design considerations. The design should stress functional simplicity, which is the approach of using minimum components to accomplish the end results. Since the three sections form an integral unit, a concept drawing should be made incorporating the final size, appearance, and functions of the robot before any detailed designs can be started.

Wheelbase Designs

At first look it is easy to assume that the wheelbase is one of the simpler sections to design. However, one must consider that it must have the structural strength necessary to support the total weight of the superstructure and manipulator plus all control electronics and all operating power supplies. If a computer is to be used to control the mobility of the robot, the selection of wheel diameters and wheel motors becomes very important. For an example, without actual computer control of the robot's mobility, the wheel motors can be reduction gear simple dc motors and the travel speed of the robot controlled either by motor gear shaft rpm or wheel diameter. On the other hand, for computer-controlled wheel drives, stepper motors should be used to drive the wheels, and the wheels must travel a given distance in one complete revolution in order for the computer to know where the robot is at any given time. A wheel with a 3.828125-in. ($3\frac{53}{64}$-in.) diameter will travel 12 in. in one complete revolution with an error factor of only 0.004 in. Forward speed would be controlled by the frequency of driving pulses to the stepper motor. As an example, a 7.5° per step motor requires 48 driving pulses to complete a single 360° movement. Thus the 3.828-in. wheel would move the robot 12 in. in the period of time it takes the computer to transmit the 48 pulses. Subsequently, a driving pulse frequency of 48 pulses per second would result in a travel of 60 ft/min.

Figure 10.10 illustrates the box design wheelbase used with the robot shown in the computer-controlled robot section in this chapter. This wheelbase design is very strong and capable of containing the rechargeable lead-acid-battery wheel motors, wheel relays, a dc-to-ac inverter, computer power supply, and battery power with ON/OFF and recharging switches. Its two wheels plus swivel caster is a simple method for maintaining stability and steering. Its disadvantages are a complex fabrication method, an excessive number of drilled holes in panels and angle stock, the amount of metal work, its excessive weight, and difficulty in finding the right wheel diameter to level the base with the swivel caster. Further, the height of the bottom of the wheelbase was such that heel or toe sensors proved difficult to install.

A superior design is shown in Figure 10.11. Here a single metal plate serves as support for everything included in the box structure in Figure 10.10. In addition, ground clearance is adjustable by altering the wheel motor mounting design. While the drawing repeats the two wheel/caster configurations of the box base, the wheels can be center mounted and by using two casters, stability is maintained. If desired, the mobility could also be accomplished by a single drive wheel coupled to a stepper motor steering gear at the position now indicated for a swivel caster and replacing the original two wheels with two casters to maintain base stability.

The wheelbase shown in Figure 10.11 can be used for circular body robots of diameter 12 to 19 in. The base must be a minimum of 0.125 in. thick for body diameters under 16 in. or bodies under 24 in. tall. Robots over these limits should have a wheelbase stock thickness no smaller than 0.250 in. A top ring or bulkhead, cut the same diameter as the wheelbase, is joined to the wheelbase by four 0.50-in.-diameter aluminum tubes to complete the superstructure. The top bulkhead does not require the weight support of the wheelbase and therefore may be cut from aluminum sheet stock 0.0625 to 0.125 in. thick. It might be desirable to cut a large circle or square opening in the top bulkhead for access to interior circuits. Once the top and bottom bulkheads (wheelbase) have been joined with the aluminum tubes, sheet aluminum, plastic, or Formica skins can be attached to the superstructure and the project begins to look like a robot.

The wheel motor base will require three cutouts, one for each wheel and the stabilizer swivel caster. No dimensions are shown on the figure due to the variations in available sizes of casters and wheels. In this case, the rule of thumb is to cut the opening 0.125 in. wider than the wheel's width and 0.125 in. longer than the maximum diameter of the wheel. The caster open-

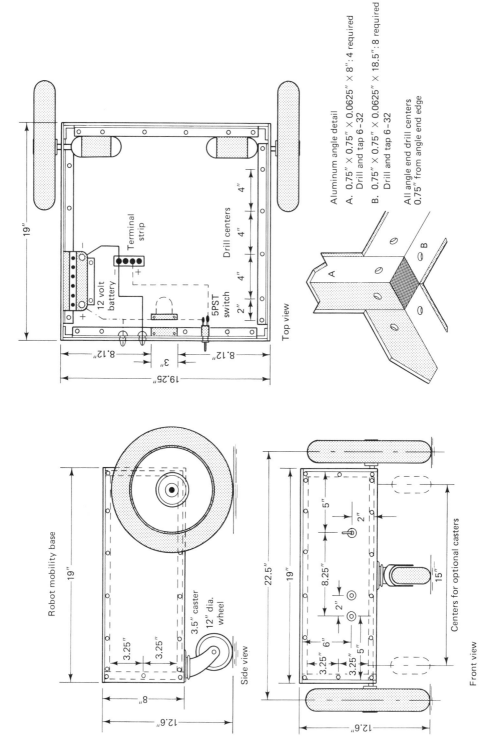

Figure 10.10 Box design wheelbase or robot mobility base.

232

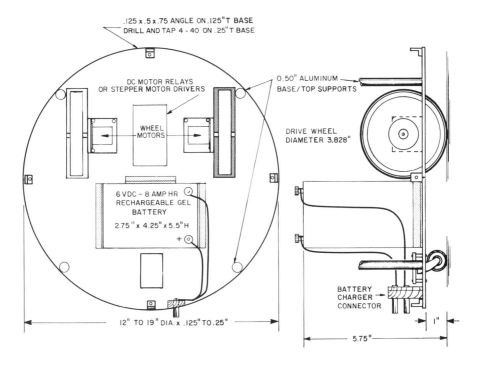

Figure 10.11   Improved design for robot wheelbase.

ing should be large enough to pass the caster completely through the opening, stopped only by the caster's mounting flange. The caster can then be adjusted to the desired bottom clearance with appropriate-length metal hex or round spacing rods which have a minimum through-tapped center for 6-32 to 8-32 machine screws; the larger screw size is preferred for the larger robots.

The wheel motors, either stepper motors or dc motors, should be mounted to an aluminum or steel angle with the motor shaft positioned through the angle plate at a height above the base to maintain the bottom clearance set by the swivel caster. The angle plate base must be attached to the wheelbase with 8-32 or 12-24 machine screws before mounting the drive motors. Dc motor-reversing relays or stepper-motor driving circuits should be mounted in the rectangular area between the two wheel motor mounts.

A rechargeable solid-gel battery is recommended as a safety measure because it not only eliminates electrolyte spillage common to lead-acid batteries, but also eliminates the generation of explosive hydrogen gas during the recharging cycle. The battery position, shown in Figure 10.11, should be boxed in with a cut-to-size $\frac{1}{2}$-in. aluminum angle mounted to the wheelbase plate. This will prevent horizontal movement, and vertical movement can be inhibited by a nylon or fiberglass strap across the battery's top. No dimensions are listed for the aluminum angle boxing material because of battery dimension variations between manufacturers.

Robot Manipulators and End Effectors

As with Cincinnati Milacron industrial robots (Chapter 8), a manipulator can have up to six axial movements. A seventh movement can be accomplished by providing one of the manipulator's sections with telescoping action. The axial movements of roll, yaw, and pitch generally take place in the end effector attached to the manipulator arm.

Figure 10.12 shows several suggested methods for making manipulator couplings to a shoulder gear or stepper motor as well as three styles of manipulator construction methods. Consideration of manipulator weight design is as important for laboratory demonstration robots as it is for industrial robots. Where the manipulator has complex axial movements, the weight factor may require counterbalancing of the manipulator in order to function with the limited power of gear or stepper motors.

Figure 10.12 includes plastic flange and ball shoulder motor manipulator attachments. Most shoulder motors do not have long shafts, and it becomes necessary to make a shaft coupler to enable the center of the manipulator to extend from the robot's body at a contact-free working distance. The adapter is easily made by drilling a 0.250-in. hole in a $\frac{1}{2}$-in. aluminum rod and a second hole at right angles to the shaft hole with a number 29 drill bit, the tap drill for the 8-32 Allen head setscrew used to lock the adapter to the shoulder motor shaft.

The number of different sizes of thin-glass Christmas tree balls and the ease with which casting acrylic resin can be worked make round ball pivots and couplings ideal for project robots. For example, to attach a shoulder motor shaft adapter to a cast acrylic ball, drill a 0.50-in hole three-quarters of the way through the ball (all the way through if the adapter is long enough). The hole will have a whitish opaque appearance due to the roughness of the drill bit. Put a few drops of a solvent-type acrylic cement in the hole, rotating the ball to coat all inside surfaces to the point where the white appearance disappears. Now force the shaft adapter into the 0.50-in. hole as far as it will go. Once the acrylic cement evaporates, the shaft should appear to be embedded in the plastic and quite secure in the ball. However, to prevent the manipulator and end effector's weight from breaking the plastic bond to the adapter, it is recommended that a 0.125-in hole be drilled through the ball and adapter shaft (this hole can go completely through the ball) and repeat the solvent cement procedure, then insert a length of 0.123-in. stainless spring steel wire. When this is done, the only way the adapter could move would be through a pressure great enough to shatter the acrylic ball.

Figure 10.12 shows a claw-gripper end effector which is closed by a

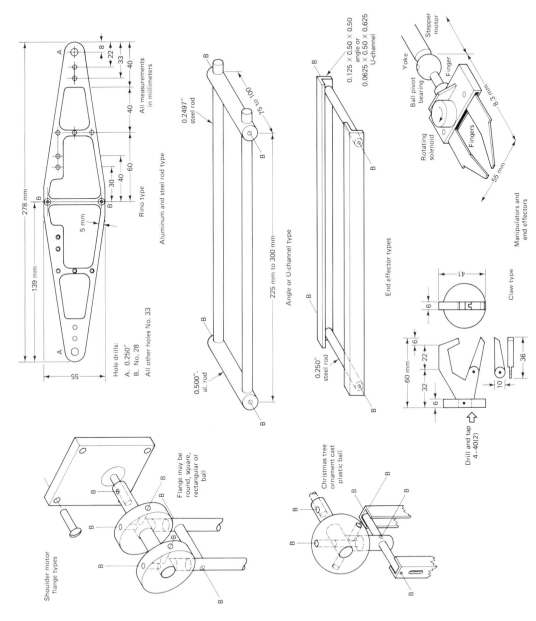

Figure 10.12 Design considerations for manipulators and end effectors.

All measurements in millimeters

Rino type

Aluminum and steel rod type

Hole drills:
A. 0.250"
B. No. 28
All other holes No. 33

Angle or U-channel type

End effector types

Manipulators and end effectors

Shoulder motor flange types

Flange may be round, square, rectangular or ball

Christmas tree ornament cast plastic ball

Claw type

Drill and tap 4–40(2)

235

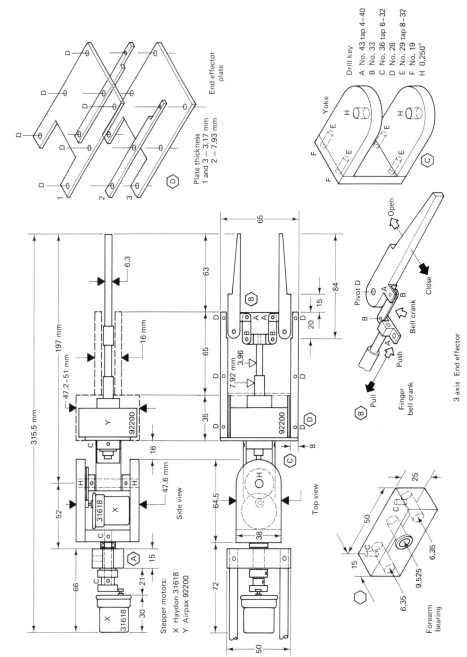

Plate thickness
1 and 3 — 3.17 mm
2 — 7.93 mm

End effector plate

Drill key
A No. 43 tap 4–40
B No. 33
C No. 36 tap 6–32
D No. 28
E No. 29 tap 8–32
F No. 19
H 0.250"

Yoke

Stepper motors:
X Haydon 31618
Y Airpax 92200

Side view

Top view

Forearm bearing

3 axis   End effector

Finger bell crank

Bell crank

Open

Close

Push

Pull

Pivot D

Figure 10.13   Design of a 3-axis end effector.

solenoid pulled wire that runs through the back of the claw to the movable finger. More in line with the learning process in mechanical engineering is the three-axis end effector shown in Figure 10.13, which shows all dimensions in millimeters. The overall length of 249 mm that the end effector extends beyond the manipulator's end gives the working size of the gripper end effector to be approximately 9.8 in.

This gripper end effector was originally designed for a rod-type manipulator with a width at the end of approximately 2 in. In order that a stepper motor fit between the forearm rods, the Haydon 31618 stepper motor was selected. A second 31618 stepper motor is used in the yoke that performs both pitch and yaw motions: yaw when the metal top of the yoke is vertical, pitch when it is horizontal. The push–pull action required to open and close the gripper fingers required an Airpax 92200 linear stepper motor. While this motor has a maximum of 3-in. "push" travel, the limitation in space between the end of the gripper and the yoke does not permit full threaded rod travel. Actually, the bell-crank movement to fully open or close the gripper fingers is only 20 mm. However, the finger pressure when fully closed is considerable and caution must be taken to prevent damage to the fingers when in the process of closing to maximum.

Techniques and design improvements now produce manipulators that are very light in weight and have almost three times the load capacity of the designs presented here. However, one must master the state of the art in any field before significant improvements can be made. These robot components can be made to machine-shop precision and quality by proper planning and using simple hacksaws, files, and an electric drill.

PROJECT ROBOTS WITH ON-BOARD COMPUTERS

Figure 10.14 shows the original prototype robot on which Unicorn-One's design was based. It employed the box-type wheelbase to allow the full body interior to be used for radio control circuits and later to house its on-board computer. Note the right manipulator; it was designed to give the upper arm section of the manipulator telescoping action. This resulted in a condition where the manipulator's overall weight and longer extension, so far off-center with the shoulder motor axis, prevented the motor from raising the manipulator until a counterbalanced weight of 15 lb was installed. The first attempt to control this robot by a computer used a 20-ft flat ribbon umbilical cable to interface an Apple II via its parallel I/O port to the on-board logic signal, control driver, and circuits. To have single key control, ASCII 7-bit codes and a latch circuit were used to energize multiple functions, and the ASCII "Null" key reset all active circuits to zero to halt the control functions. This method of programming the robot worked much like the first Altair 8080b computer, which required manual setting of 8 data switches and 15 addresses

Figure 10.14 Original prototype robot at the development stage of computer control via an Apple parallel I/O and umbilical cable.

switched to program; it was very slow and there was no way to see what had been entered. To achieve independent control of each of the robot's function, it was modified for control by pulse-encoded AM radio signals. Figure 10.15 shows the bank of eight servomotors which activated control relays by having the servo bell crank operate the toggle arm of a single-pole,

Figure 10.15 Servo motors and switch bank required for robot radio control.

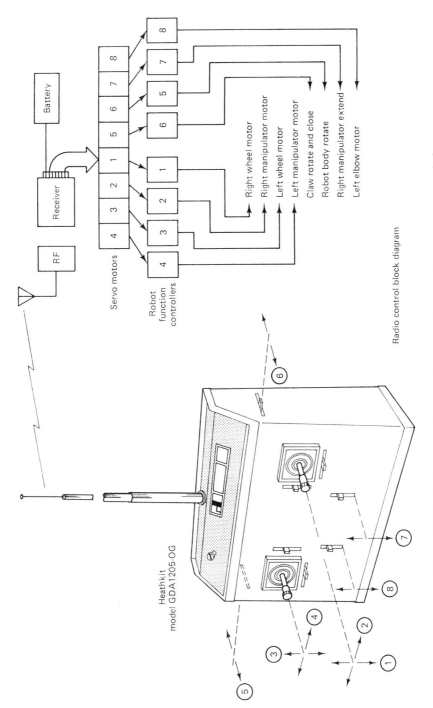

Battery

Receiver

RF

Servo motors

Robot function controllers

Right wheel motor
Right manipulator motor
Left wheel motor
Left manipulator motor
Claw rotate and close
Robot body rotate
Right manipulator extend
Left elbow motor

Radio control block diagram

Heathkit model GDA1205-OG

Figure 10.16 Radio control transmitter operating control movements and the robot functions controlled.

double-throw, center-off switch (SPDP-CO). This method enables the switch to apply energizing voltage to the relay coils of two double-pole, single-throw relays cross-wired in parallel. In one position of the switch, the relay contact pair would apply the voltage polarity to a motor for clockwise rotation. In the opposite switch position, the second relay receives the coil energizing voltage, and the contact pair reverses the polarity of voltage to the motor and the motor runs in a counterclockwise direction. Figure 10.16 shows the radio control transmitter's eight control movements for servomotor control and the robot functions that each servomotor activates. The dual relay wiring schematic is shown in Figure 10.17 and the cross wiring necessary to reverse voltage polarity is shown in the photograph of a wired dual relay in Figure 10.18.

The robot was modified from radio control to on-board computer con-

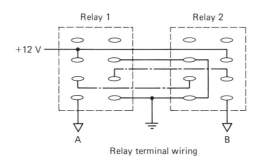

Relay terminal wiring

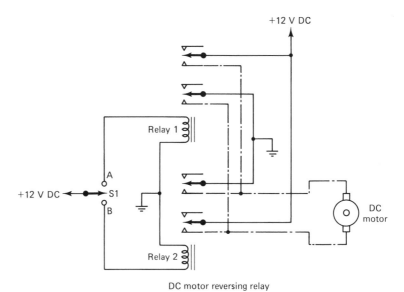

DC motor reversing relay

Figure 10.17   DC motor reversing relay wiring diagram.

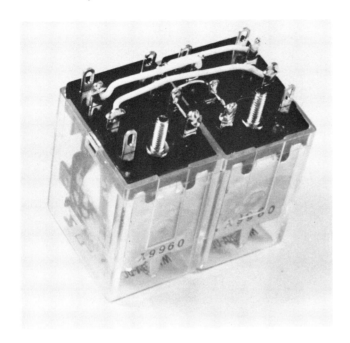

Figure 10.18 Wired dual relay used to reverse dc motor rotation direction.

trol by removing all radio-control servomotors, switches, and so on. Figure 10.19 shows the robot and its on-board computer: a Wintek 6800 modular computer using a Motorola 6800 microprocessor. The figure also shows the four parallel I/O ports that carried the computer control digital signals to new digital logic input circuits to drive the robot functions.

The wheel support box was modified by the addition of an inverter power unit which changes 12 V dc to 120 V ac in order that the Wintek computer power supply could be operated from the robot's on-board lead-acid battery (see Figure 10.20). The Wintek computer could be programmed by an on-board hexidecimal keypad, cassette tape, or by a CRT terminal via a RS-232 I/O port. When the on-board computer was programmed from the terminal or program loaded from cassette, the program required starting with a time-loop delay to allow time to disconnect the RS-232 cable after the RUN command had been entered.

The development of this robot was begun in June 1978 and converted to radio control in early 1979. By June 1980 the radio control was replaced with the Wintek computer. This time period suggests that the robot might be one of the first computer-controlled model robots, certainly one of the first with 16k on-board RAM. However, its computer operation was short lived because it was acquired by the Instituto Politecnico Nacional in Mexico City. At their request, the computer was removed and the robot was returned to its radio-control state to become a robot chess player.

Figure 10.19 The original prototype robot and its onboard Wintek Modular 6800 computer.

Figure 10.20 Robot wheelbase box showing inverter and Wintek computer power supply.

Single-board Computers for Robots

The size of computers suitable for model robot control has been condensed drastically from the six printed circuit cards and three voltage power supplies of the Wintek computer to a single card half the size of a Wintek card. One of the first available single-board computers was the Technical Micro Systems BASYS/1 computer. This computer, shown in Figure 10.21, was programmed in an eighth control language (a modified Tiny BASIC) which offered a number of subroutines in its ROM. It has 4k on-board RAM, a parallel I/O port, and an RS-232 serial port with a 20-mA current-loop optional input/output. One unusual feature is the user breadboard area on the left side. The total size of its 22/44 edgecard connector circuit board is 4.5 in. × 6.5 in. without the external power supply. Operating voltage re-

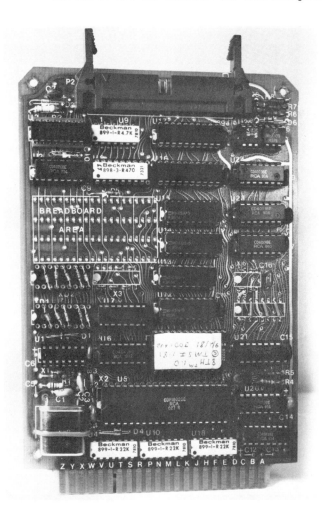

Figure 10.21 Technical Micro System's BASYS/1 single board microcomputer.

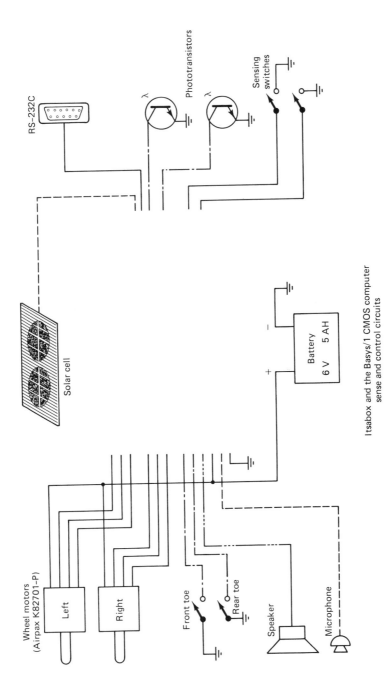

Figure 10.22 External controlled capabilities of the BASYS/1 computer.

Itsabox and the Basys/1 CMOS computer
sense and control circuits

244

quirements are +5 V dc from ac power supply or battery. The BASYS/1 computer was combined with a box having two servomotor-driven wheels to become the first on-board computer-controlled robot kit, called Itsabox. Figure 10.22 shows in schematic/pictorial form the sense and control circuits of the BASYS/1 and the Itsabox computer kit.

The BASYS/1 single-board computer was the on-board computer for the computerized Apple logo robot shown in the schematic, Figure 10.23. Its small size limited the number of controlled functions but it was adequate to perform its program. Using the Itsabox demonstration program, the Apple logo robot, once programmed, could be placed on a large table where it immediately started exploring, emitting computerlike noises from its internal speaker. Upon encountering an edge of the table, its toe or heel sensor would halt the robot, it would shout "Woops!," back up, turn, and dash off on its new heading until it again encountered a table edge. Its demonstration program looped continuously, so once started, it would continue the amusing antics until either turn-off or the power supply became exhausted. Figure 10.24 shows the BASYS/1 computer installed inside the Apple logo robot. The wheel servomotors are located on both sides of the computer board, near the middle, and the user breadboard section was used to interface the computer to the relay driving transistor circuits under the computer board. Figure 10.25 is a schematic of the stepper-motor control circuits and

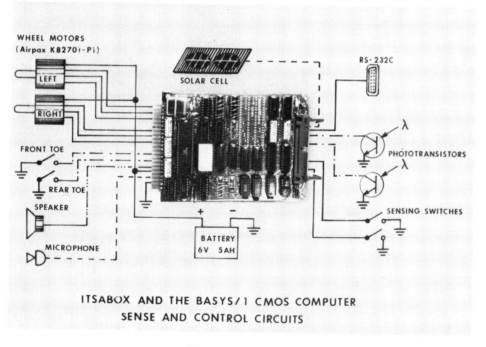

ITSABOX AND THE BASYS/1 CMOS COMPUTER
SENSE AND CONTROL CIRCUITS

Figure 10.22 (cont.)

connector pin numbers from which running pulses are taken. The figure also shows the schematic and connector numbers for the robot's speaker, sensors, power, and RS-232 input. The Apple logo robot is now displayed at the Apple Computer Company in Cupertino, California.

Latest State-of-the-Art Single-board Computers

CMOS technology has been advancing so rapidly that the chapter on CMOS technology (Chapter 5) has been rendered virtually obsolete. It was pointed out that the original CMOS inverter occupied a space 125 μm across and later this had been reduced to only 65 μm and gave superior performance

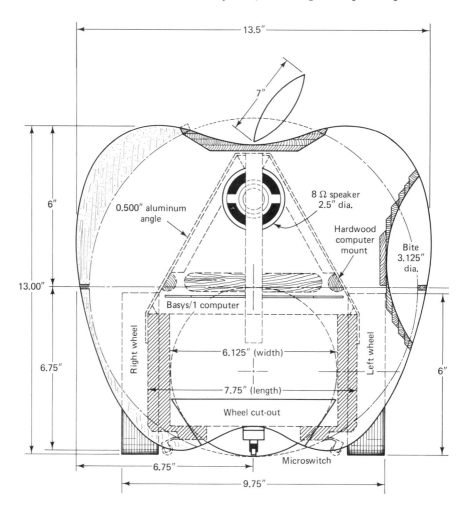

Figure 10.23 Schematic drawing of the computerized Apple logo robot structure.

and reduced process time. Current R&D work in CMOS technology predicts that the 65-micron area will shrink to less than 1 μm by 1989 or 1990. It is further predicted that 1-megabit RAM chips will appear as early as 1987 and 4-megabit memory chips will be in use by 1992.

As a measure of how advances in CMOS technology can have an effect on single-board computers, compare the circuit board of the BASYS/1 in Figure 10.21 with the Mini-73 in Figure 10.26. The BASYS/1 contains 20 integrated circuits, including ROM, RAM, and the microprocessor. The Mini-73 has 14 integrated circuits, including the microprocessor and a bidirectional parallel I/O. Where's the ROM? The microprocessor contains 4k ROM internally and user option CMOS RAM is available to 8k. There is a new note to single-board computers, the CMOS technology chip's current drain is so low that the newer circuit boards come equipped with rechargeable-battery-power backup.

It is difficult to comprehend exactly what the CMOS technology is accomplishing in microelectronics. As a matter of fact, few can conceive a dimension given in microns. In decimal inches, 1 μm is equal to 0.000039 in. or approximately 40 millionths of an inch. Objects of this size require a high-powered microscope to be seen by the human eye. To have some idea of the marvel of microelectronics, look at your little fingernail. It is approximately

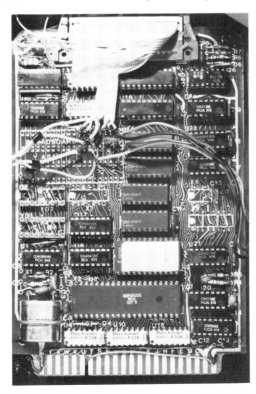

Figure 10.24   BASYS/1 computer mounted inside the bottom half of the Apple logo robot.

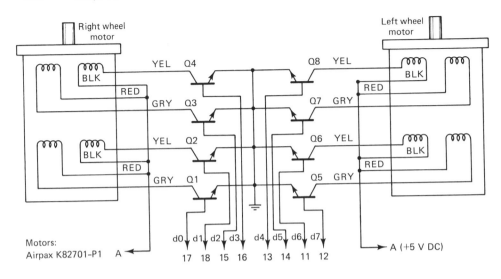

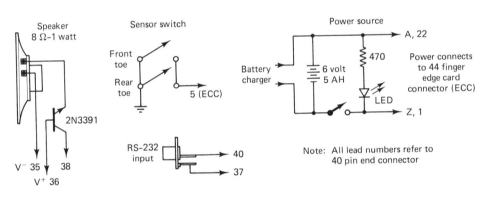

Figure 10.25 Schematic diagram of the BASYS/1 control circuits in the Apple logo robot.

3/8 in. wide and 7/16 in. long. Now look at Figure 10.27 and try to visualize the design printed on your fingernail. Figure 10.27 is a photomicrograph of the National Semiconductor NSC800 microprocessor, which contains the arithmetic-logic unit (ALU), 14 general-purpose registers, 2 index registers, program counter, stack pointer, refresh register, interrupt register, clock, timing, and sequencing circuits. In addition, it has the capability of addressing 64k bytes of memory and 256 I/O ports.

Selection of a training-aid computer for electromechanical engineering courses involves careful evaluation of available computer capabilities and the feasibility of upgrading the basic system at a future date. Computers are comparable to any commercial product; they are manufactured to meet the

Figure 10.26 Mini-73 single board computer with battery "keep alive" memory.

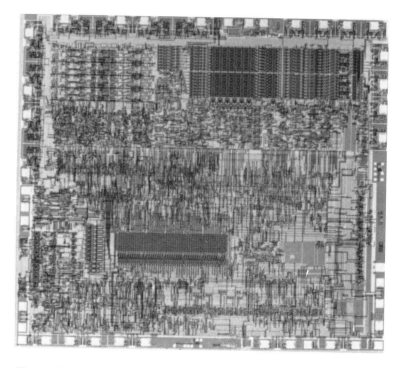

Figure 10.27 Photomicrograph of the National Semiconductor NSC-800 microprocessor.

consumer market in existence 3 to 5 years ago. This is significant improvement over the 7- to 10-year time lag between prototype and commercial production that was common 20 years ago. It is not surprising that data processing computer production is second to none and control computers at the very bottom level. Those in the manufacture of computer-controlled machines or processes are forced to develop their own specialized control computers or rely on modifications to available computers. Naturally, computers manufactured for a specific use seldom have available internal space for specialized modifications and finding compatible modification-required interfaces was even more difficult. Fortunately for schools and colleges, the computer industry is beginning to wake up where specialized computers are concerned and are starting to offer the much needed product lines.

THE NATIONAL SEMICONDUCTOR INDUSTRIAL MICROCOMPUTER SYSTEM

One such firm is National Semiconductor, which has developed a new line of CMOS Industrial Microcomputer (CIM) board-level products. Figure 10.28 shows the NSC800 microprocessor board of the CIM family of computer peripheral interfaces. Each board has the same size configuration:

Length	6.30 in. (160 mm)
Width	3.94 in. (100 mm)
Height	0.50 in. (13 mm)
Weight	1.40 oz (36 g)

The microprocessor is the NSC800-based P(2)CMOS microprocessor, which operated with a system bus called CIMBus which was specifically designed for control-oriented run-forever applications. A distributed input/output bus (DIB) is an isolated I/O interface provided the CIM line which was designed to provide a flexible error-free interface to remove devices.

The CIM central processing unit (CPU) is available in three models:

CIM-801	1-MHz NSC800
CIM-802	2-MHz NSC800
CIM-803	4-MHz NSC800

Each CPU has a capacity of 2k to 4k PROM, 2k RAM, two 16-bit counter/timers, eight vectored interrupts, 16 discrete I/O lines, and a power-saver mode. The CIM-800 CPUs are supported by add-on boards as follows:

Memory	Six RAM expansion boards for 8k, 16k, 32k, 64k, and two boards without memory chips installed

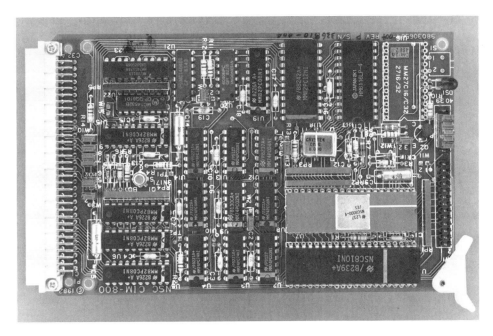

Figure 10.28 National Semiconductor NSC800 microprocessor CIM-Bus central processing unit.

Digital I/O	Six boards, including single and double serial I/O, fiber optic interface, parallel I/O, frequency/period measuring board, and PWM output board
Analog I/O	Three 12-bit analog boards, including two inputs with gain or 20 or 100 and one analog output board
DIB boards	Two DIB (distributed input/output bus) I/O boards, one for DIP interface and one for power
Controller boards	Four clock/calendar boards, two math processing boards, and a floppy disk board
Ancillary boards	Sixteen boards, including card cage and back planes, voltage regulators, battery charger, extenders, prototyping boards, disk, printer, and serial cables, firmware monitor, and EPROM programmer

The National Semiconductor CIM-800 CPU is further supported by the Starplex Development System with an optional CP/M operating system, and the upgraded Starplex II System, which offers a real-time in-system emulation and high-level languages, including PL/M, Pascal, BASIC, and Fortran. The total CIM-800 system, including Starplex or Starplex II, results in a laboratory computer system of appropriate size, adequate interface capability, program language flexibility, and off-the-shelf software for computer-controlled teaching-aid robot projects.

CHAPTER REVIEW QUESTIONS

1. What mechanical motions can be demonstrated by a robot?

2. What purpose does the Turtle's plastic dome perform?

3. What type of energy does the Line Tracer robot use to sense the black line?

4. Since the control whistle of the Piper Mouse robot is not in the ultrasonic range, would a loud noise be able to activate the control sensor?

5. The Memocon robot has 4k RAM to store control functions. Does it have a microprocessor?

6. What are the two programs that enable the RBX5 robot to learn from its experience?

7. What is the range of the RBX5 robot's sonar detector?

8. What does the name Unicorn represent as related to the robot Unicorn-One?

9. Two examples of controlled figures were given in the chapter and referred to as robots. What would be necessary to refer appropriately to a model electric train as a robot?

10. To what section of a robot does the term *superstructure* refer?

11. What is functional simplicity?

12. How would you determine the diameter of a stepper-motor-driven wheel if it were to travel 2 ft in one 360° rotation of the stepper motor?

13. What is the purpose of a part containing a flange?

14. What two safety factors are associated with rechargeable gel batteries over lead-acid batteries?

15. A working robot manipulator generally functions over six axes. What function would constitute the seventh axis?

16. What must be added to the opposite side of the manipulator shoulder junction to offset the manipulator's weight?

17. What is the function of a bell crank?

18. Programming a robot's on-board computer requires a time loop at the start of the program. What purpose does the time loop perform?

19. What program language was used to program the Apple logo robot?

20. What single accomplishment in CMOS technology will be instrumental in producing 1-MHz RAM chips in the next few years?

21. Name three major advantages of the National Semiconductor NSC800 microprocessor system over other current computer systems.

GLOSSARY

AM. Amplitude modulated.

Ancillary. An accessory component essential to the operation of related peripherals.

ASCII. American Symbolic Code for Information Interchange.

Bell crank. A bar or L-shaped lever used to transfer motion from one direction to another (e.g., a push/pull forward motion changed to a push/pull horizontal motion).

Breadboard. An insulated section or material on which electronic circuits may be assembled for test or trial use without short circuits affecting the operation of the test or trial circuit.

CIM. Control industrial microprocessor.

Cimbus. A system's bus for the NSC800 CMOS microprocessor, designed especially for control-application central processing units.

CPU. Central processing unit.

DIB. Distributed input/output bus.

Edgecard contacts. A series of fingerlike electrical contacts on one or both sides of a circuit board that make electrical contacts to external power or circuits when inserted into an appropriate mating connector.

Flange. A surface surrounding or part of a component used to attach the component to another device or component; a projecting rim or collar to provide a supporting area for attachment to another device.

Hexidecimal. A number system to the base 16, which contains digits 0 to 9 and letters A to F to distinguish the number system from a total decimal base 10.

Robot. A self-contained mechanical device having the capability of receiving instructions and carrying out the instructions through the direction of an on-board computer; a computer-guided mechanical device with the ability to perform routine tasks and movements on its own power.

RPM. Revolutions per minute.

Umbilical cable. Joined together by a bundle or cable of wires over which life- or operation-supporting energy is carried (e.g., power or data operational signal carrier between computer and robot).

11

University Research Robot Projects

The approach to using model robots as training aids, the subject of Chapter 10, was based on the number of research projects conducted at universities since 1967. These projects produced a variety of robots and control methods which became the subject of master's theses and technical reports. It is appropriate that a summary of some of the outstanding projects be included as precedence for robotic projects at the college and university level.

SHAKEY (1967-1969)

One of the earliest projects combining a robot with artificial intelligence was the robot Shakey at the Stanford Research Institute in 1967 [Nilsson, 69; Coles, 69]. This automation was linked to an SDS 940 time-sharing computer via radio and programmed in FORTRAN.

Shakey's main sensor was a rotatable camera which required multilevel software to achieve visual inputs in order to navigate around the laboratory. Shakey could be programmed in simple English commands and it would parse the sentence and call up the appropriate FORTRAN program to carry out the command.

In appearance, Shakey employed a wheelbase not unlike the wheelbase box in Chapter 10. Its computer interface, logic circuits, and camera control functions were contained in a square "body" and its rotatable camera perched on top of the control box like a head. A contact sensor ring, around the top of the wheelbase box, served as a bump detector when encountering path obstacles.

The integration of the hierarchical levels of software gave Shakey the reputation of being the state of the art in robots for many years. However, from the standpoint of performing as a completely autonomous robot, Shakey was a failure.

THE STANFORD CART (1973-1981)

In the early 1970s, Hans Moravec began the development of a remotely controlled video-camera-equipped mobile robot at the Stanford University Artificial Intelligence Laboratory [Moravec, 81, 83]. A simple four-wheel cart served to support the vision system, which enabled the cart to maneuver through cluttered areas through the aid of appropriate navigation and obstacle-avoidance software.

Instead of a radio link between the robot, as was employed with the Shakey robot, the cart used a television cable link between the cart and the image processing computer. The video camera was mounted on a rail and could be moved to nine different positions by remote control. In this manner nine different images could be digitized to produce a three-dimensional

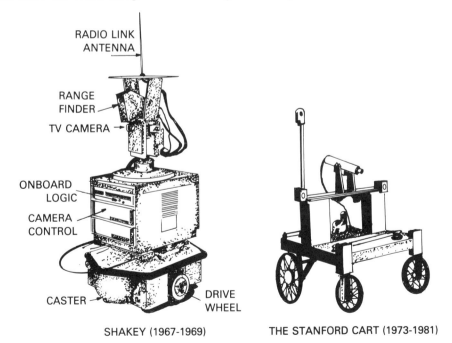

SHAKEY (1967-1969)   THE STANFORD CART (1973-1981)

Figure 11.1   Shakey and the Stanford cart.

picture of any obstacle in the path of the cart. Obstacles were detected by scanning each image and extracting its key features. Once all nine pictures had been scanned for key features, a correlator routine would compare change in feature pixel position between each picture. The video images are taken and fed to the computer while the robot is in motion, and the distance between the obstacle and the robot is determined by analysis of the change in image pixel position to the distance traveled by the robot during the image processing time. Artist drawings of the configurations of the Shakey robot and the Stanford cart are shown in Figure 11.1.

SCIMR (1981)

A self-contained independent mobile robot (SCIMR) was built at the University of Pennsylvania Moore School of Electrical Engineering by R. L. Anderson in 1981 [Anderson, 81, 82]. It was basically a two-wheel-and-caster wheelbase with an on-board computer: three interconnected 6808 microprocessors with 4k RAM-controlled separate functions. One microprocessor controlled the wheelbase's two rear wheels to provide motion and steerage, one controlled a rotatable Polaroid ultrasonic transducer, and one coordinated all functions withe the first two microprocessors.

The hallways of the Moore School of Electrical Engineering served as the navigational laboratory for SCIMR. Programming was simplified since the robot dealt only with passageways and intersections. The servomotor-rotated sonic transducer takes a measurement perpendicular to the robot, which establishes its position with respect to the hall wall. Additional measurements are taken forward and to the rear of the robot. The input data from the sonar sensor then enabled the robot to perform steering corrections to maintain its distance from the wall as it moved down the hall. On reaching an intersection, if its sonar detects no return signal from the wall, it was programmed to make a 90° turn and enter a new path down the next hall. SCIMR's only problem was in the 20° beam width of the sonar transponder, which frequently gave misleading data from signals received from open doors, cabinets, or chairs along the hall wall, which caused false course corrections and erratic navigational paths.

THE UNIMATION ROVER (1983)

Unimation, Inc., one of the leading manufacturers of industrial robots, recently attempted to build a robot manipulator on a mobile base. This project has since been passed on to the Mechanical Engineering Department of Stanford University. The intent is to make the Unimation Puma robot mobile by the addition of omnidirectional wheels originally developed for the Veterans' Administration Rehabilitation Research System (see Chapter 12).

The Puma's motor controllers have large power consumption and with six batteries, the operational time is an hour or less. Wheel-drive motors must have the power to transport not only the overall weight of the Puma robot but the additional weight of the batteries necessary for powering the Puma and its mobility base.

It is apparent that robotics is a systems problem and much more goes into making an intelligent mobile robot than in combining a number of sub-systems that works well independently. This is not to suggest that it is impossible to make a totally autonomous robot, as the feature robots, Robart I and Robart II, will show.

ROBART I (1980–1983)

Robart I was designed and built by H. R. (Bart) Everett at the Naval Postgraduate School as a feasibility demonstration of an autonomous robot. It is the first known mobile sentry robot to be constructed and the subject of a master's thesis [Everett, 82a, 82b]. Robart I, shown in Figure 11.2, can be

Figure 11.2   Robart I (1980–1983). (Courtesy of LCDR H.R. Everett.)

TABLE 11.1   Specifications: Sentry Robot—Robart I

Height	62 in. (1574.2 m)
Width	15 in. (381 mm)
Length	26 in. (660.4 mm)
Weight	80 lb (36.288 kg)
CPU	6502 Synertex System SYM-1; single-board computer
Memory	4k RAM on-board
	32 k RAM expansion board
	20k ROM on-board
	2 Shugart $5\frac{1}{4}$-in. double-density disk drives
	Cassette drive
Power	12-V 20-Ah lead-acid battery
Drive	Tricycle wheelbase—tandem drive to steerable front wheel. Maximum turn angle 80° left or right, turn radius 12 in.
Collision avoidance	Long-range sonar and positionable near-infrared short-range and near-infrared proximity detectors, tactile sensors, and drive overload monitoring
Intrusion detection	True infrared body heat sensor with 50-ft range, near-infrared active sensor, ultrasonic sensor, three optical motion detectors, discriminatory audio detection
Security-related functions	Smoke and fire detection, flooding detection, earthquake detection, severe weather condition detection

programmed to patrol any area sensing for fire, smoke, toxic gas, intrusion, natural disasters, or severe weather. On detecting any of these hazardous conditions, Robart I's programming enables it to give appropriate warnings.

The previously mentioned research or thesis robots were basically little more than mobility bases supporting optical or contact sensors. Robart I was the first thesis robot to have multiple sensors and fully controlled by its on-board computer. Table 11.1 lists the unusual specifications of Robart I.

The primary purpose of Robart I was to serve as a mobile platform for research and experimentation in the areas of artificial intelligence, computer interface techniques, speech synthesis, recognition, and to resolve the associated problems of electrical and mechanical design.

Robart I Design Considerations

1. Design a suitable chassis and body framework with a suitable mobility drive with steering capability.
2. Design a rotating head assembly to be mounted on top of the body trunk and have positioning capability of 100° movement left or right of center position.
3. Develop speech synthesis for audible warnings.

4. Develop an optical photocell array in the head section to locate tracking beacon on battery-recharging station.

5. Develop a single-transducer sonar system for range detection of immediate path obstacles.

6. Develop a multielement near-infrared collision system for object detection in the first and fourth quadrants relative to the centerline.

7. Develop physical contact bumpers or feelers for collision detection.

8. Develop sensor systems for intrusion detection.

9. Development of complete operating software with printer hard-copy provisions.

10. Develop software provisions to enable operator to request control from the CPU when troubleshooting or to request a specific behavior pattern subroutine.

Selecting the size of the robot required special consideration of several design requirements:

1. The optical sensor used to seek out the battery-recharging station must be high enough to "see" over most obstructions.

2. Maximum body trunk height must be within the limits of 6-ft-standard-length aluminum angle.

3. Body trunk must be tall enough to locate photocell sense array at an height of 44 in.

4. Base width must be wide enough to provide room for the two drive motors attached to either side of the front steering wheel and wide enough to provide a stable foundation for the body trunk.

The final design set the base length at 25 in., which proved to be very stable yet narrow enough to maneuver through passageways of almost every size. Referring back to Figure 11.2, note how the rear wheels are recessed into the sides of the rectangular base. This allowed the side panels of the base to present a smooth surface that was relatively free of any protrusions. The tricycle wheelbase was selected primarily for comparison to a preliminary prototype base which had separately controlled driven wheels at the rear and a swivel-type caster in front to maintain stable mobility. The steerable front wheel offered a slight advantage in control of steering over control of the two driven-wheel motors as the steering method.

Microprocessor Selection

The control supervisory computer was the SYN-1 single-board computer using the 6502 microprocessor. It was manufactured by Syntek, Inc., Santa Clara, California. It was chosen as the prototype computer because of the extensive hardware contained on a single board, its superb assembler/editor for software development, and its relative low cost.

The 6502 MPU operates into a 16-bit address bus and an 8-bit data bus and accesses to 4k on-board RAM and 28k on-board ROM. An additional 32k off-board RAM is accessed through a 44-pin expansion connector. The computer board also contains three 6522 versatile interface adapters (VIAs) and one 6532 peripheral interface adapter (PIA), which together provide a total of 71 input/output (I/O) lines for control functions.

The SYM-1 computer is mounted at midheight of the body trunk, forming the heart of all electrical and electronic control. Its main function is to serve as a dedicated controller, but it can be borrowed to interface to a Synertek KTM-2 terminal by using its RS-232 Serial I/O connector. In the mode, the robot will remain motionless beside the KTM-2 terminal stand as long as the Syn-1 is under the terminal's control. During this time period, the robot receives supplemental input power over the RS-232 cable. Once released by the operator, the computer first verifies that the RS-232 circuit has been disconnected, then proceeds under its own control and battery power. Should the connecting RS-232 cable have not been disconnected, operator assistance is requested by the robot using its speech synthesis circuits.

Robart I Speech Synthesis Capability

The prototype speech synthesis was implemented with the National Semiconductor Digitalker, model DT1050, with two vocabulary instruction sets stored in EPROM, giving the robot a vocabulary of 280 words. Originally, an unlimited vocabulary created by repeated used of phonemes was considered. However, this system would impose greatly increased demands on the host microprocessor in terms of execution time and memory space over the use of a fixed vocabulary system with speech instructions in EPROM and was discarded.

The Digitalker DT1050 is interfaced to the SYM-1 computer via two parallel I/O ports, one of which supplies the EPROM starting address for instructions needed to generate a desired word. Since the EPROMs are addressed by the DT1050, it does not take up address space in the SYM-1 system. A portion of the second parallel I/O is used to initiate the speech output and to detect completion of each word. This amounts to the SYM-1 simply instructing the DT1050 to speak a particular word, then checking to see that the word has been spoken before requesting the next word.

Two subroutines, VOX-1 and VOX-2, request words from vocabulary list 1 and 2, respectively. The hex address of the desired word is loaded into the X register of the SYM-1 and then to the appropriate VOX subroutine. After the desired word is identified, Subroutine Talk is called to initiate the speech, then waits until the VIA 1's busy signal on PB7 has been cleared.

PIA Parallel Interface Layout

When the SYM-1 is used as a dedicated controller, it communicates to the outside peripherals or controlled circuits through its three 6522 VIAs and the 6532 PIA. Each interface contains two parallel 8-bit I/O ports (port A and port B), which can be used to read sensory information or to send commands to external circuitry. Each 6522 contains 16 internal 8-bit registers, as shown in Figure 11.3. The internal organization of the 6532 PIA is very similar and its differences are so insignificant that it can be considered identical to the 6522 VIA in operation. To facilitate VIA identification in the cir-

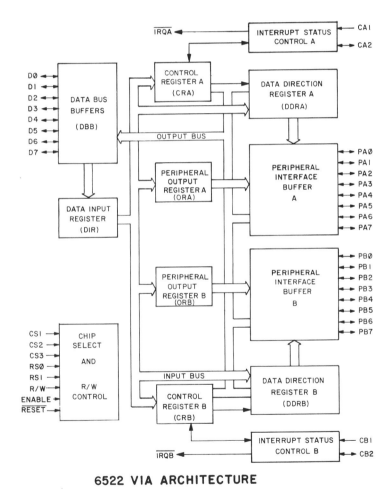

6522 VIA ARCHITECTURE

Figure 11.3 6522 versatile interface adapter (VIA) schematic. (Courtesy of LCDR H.R. Everett.)

cuit operation descriptions, the VIA related to the circuit will be identified as 6522-1, 6522-2, 6522-3, or 6522-4.

The configuration of the SYM-1 is such that all 16 VIA or PIA registers appear to the CPU as a specific memory location within its addressable 64K RAM memory space. It reads or writes data from the registers as it would from any memory location. As shown in Figure 11.3, for each A or B port of a VIA or PIA, there is an 8-bit input/output register that the CPU actually reads from or writes to instead of going directly to the A or B port. The I/O register connects directly to the 8-bit data bus via one of the expansion connectors on the SYM-1 board. In operation in the output mode, data can be loaded into an I/O register and the data lines assume the transistor-to-transistor logic (TTL) voltage level dictated by the subsequent register contents. For each bit that is high, the output line will go high, whereas each low bit will cause its associated data line to go low. These registers can be incremented, decremented, or rotated by assembly language instructions, causing the voltage levels on the output lines to react accordingly. The TTL compatibility of the PIA's output lines enables the output to interface directly to control solenoids, motors, lamps, or other sensors according to the control program.

In the input mode, the TTL voltage determines the individual bit values within the input register. This enables the computer to determine the state of any control line by simply reading the data in the input register and masking off all but the desired bit with logic operations in the accumulator.

Each PIA has two data direction registers which determine the mode of operation for each individual bit stored in the input/output registers; A or B. If the MPU writes a logic 0 into a specific bit in the data direction register's A port, that bit will be configured as an input on the input/output register's A port. By the same token, if the MPU should write a logic 1 in the data direction's A port, the bit will become configured as an output on the input/output register's A port. In this manner any line to the PIA's A or B input/output registers can be used as either an output or an input line according to the control program's instructions.

The remaining registers associated with the 6522 VIAs are used for shift register operation, event timers, and interrupt processing. These operations will not be discussed here since it is the I/O bidirectional lines for control functions that are being emphasized. For detailed operation of these registers, refer to Scanlon [80] and Zaks [79].

As the prototype design and construction progressed, it soon became apparent that the SYM-1's 71 I/O lines where insufficient due to the number of inputs and outputs needed for autonomous control of a mobile system. Thirteen I/O lines are required for communication with the speech synthesis microprocessor, eight for the head and wheel-drive position control, five more for the operator's control interface, and another 16 for tactile sensors.

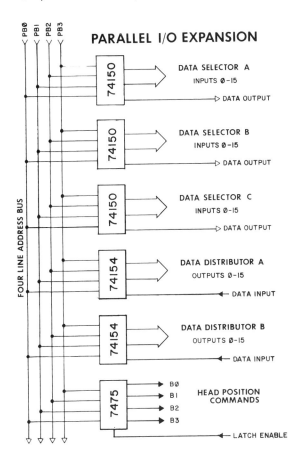

PARALLEL I/O EXPANSION

PB0 PB1 PB2 PB3

FOUR LINE ADDRESS BUS

74150 — DATA SELECTOR A INPUTS 0–15 ⊳ DATA OUTPUT

74150 — DATA SELECTOR B INPUTS 0–15 ⊳ DATA OUTPUT

74150 — DATA SELECTOR C INPUTS 0–15 ⊳ DATA OUTPUT

74154 — DATA DISTRIBUTOR A OUTPUTS 0–15 ← DATA INPUT

74154 — DATA DISTRIBUTOR B OUTPUTS 0–15 ← DATA INPUT

7475 — B0 B1 B2 B3 HEAD POSITION COMMANDS ← LATCH ENABLE

Figure 11.4   Parallel I/O
expansion board schematic.
(Courtesy of LCDR H.R. Everett.)

It then became necessary to create a new four-line address bus to serve six interface boards which connect the CPU to various sensors and outputs.

The four-line bus is driven by PB0 through PB3 of 6522-1, one of the three VIAs on the SYM-1. In turn, the four-line bus drives three 74150 16-input data selectors, two 74154 16-output data distributors, and one 7475 4-bit latch (refer to Figure 11.4). This combination provides an additional 84 I/O lines at the expense of 10 of the original 71 I/O lines on the SYM-1.

74150 Data Selector Operation

The three data selectors are driven simultaneously by the four-line address bus when the CPU sets PB0 through PB3 with a complement of the selected input appearing on the selector output, pin 10. The selector outputs are inverted before being read by the appropriate inputs on the 6522. This is necessary to maintain the same logic-level voltages as seen by the selector input.

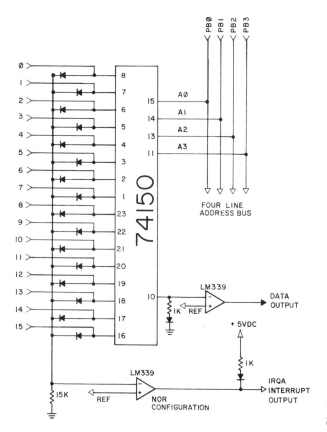

DATA SELECTOR A

Figure 11.5 National Semiconductor 74150 4-line address bus schematic. (Courtesy of LCDR H.R. Everett.)

The inversion is accomplished by a spare comparator in the LM339 quad comparator used to implement a 16-input NOR gate (refer to the schematic, Figure 11.5). The CPU must set the value of the desired input number onto the four-line address and read the appropriate selector value over the 6522 VIA's input. As an example, to read the target input used in homing into the battery-recharging station, the CPU must set the four-line address bus to 0101 to select input 5 on all address lines and read the output of data selector C over PA6 on 6522-3. Subroutines READ A, READ B, and READ C take care of manipulating address and data lines when a desired address number is loaded into the X register before an appropriate subroutine is called.

There are three data selectors and two data distributors shown on Figure 11.4. Data selector A is entirely dedicated to tactile and proximity sensors with interrupt capability. Data selector B is dedicated to handling all alarms and internal circuitry check points with its interrupt capability. Data selector C is used to read miscellaneous inputs and does not have interrupt capability.

74154 Data Distributor Operation

The two 16-output 74154 data distributors are driven by the same four-line address bus as the 74150 data selectors. However, the data distributors require three additional control lines, referred to as data 1, data 2, and data 3. These three lines are normally held in a high state. Data 1 and data 2 provide inputs to data distributors A and B to select the desired data distributor. All 16 inputs to the data distributors are normally high except the selected output, which follows the applied input. Thus if all inputs are held high, there will be no change in the outputs regardless of the status of the address lines.

When the CPU desires to momentarily pull low an output on distributor B, it first sets the appropriate binary value on the four-line address bus (0011 for output 3, for example) and momentarily sends data 2 low. This causes output 3 on distributor B to go low. However, output 3 on distributor A, although selected by the address on the four-line address bus, does not go low because it follows data 1 high or low states. It is important to note that this system can be used only to strobe outputs on selector B, because they cannot be held low for any length of time without tying up the CPU. For this reason these outputs can be used only as negative-going triggers to initiate timing sequences or actions subsequently controlled by other circuitry.

The third control line is used with distributor A to overcome the nonlatching problem of distributor B and provides a means for latching distributor A's outputs high or low. Figure 11.6 shows how each output on distributor A is used to clock a D-type flip-flop. The eight D-type flip-flops have their D inputs commonly connected to data 3. However, the only way any of the eight flip-flops can assume the logic level of data 3 is when one or more flip-flops are also clocked by distributor A. The clocking sequence requires placing the address of the selected flip-flop(s) onto the four-line address bus with data 1 and data 3 high. Data 3 is then set to the flip-flop's desired output; either high or low. Data 1 is then strobed to clock only the selected flip-flop(s), which then holds the output logic state determined by data 3. Although this appears to be a complex method of manipulating address and data control lines, it is simplified by subroutines Pin lo and Pin hi, which set the address according to the contents of the X register, then set data 3 high or low according to the desired output of the selected 7474 flip-flop. For example, to turn on Robart's spotlights, the programmer would load the X-register with hex 01, then call subroutine Pin hi. This would set flip-flop 1 to high, which then turns on the spotlights.

The final circuit on the expansion I/O board is the 7475 quad latch used to store the 4-bit commands for head positioning. The latch is level

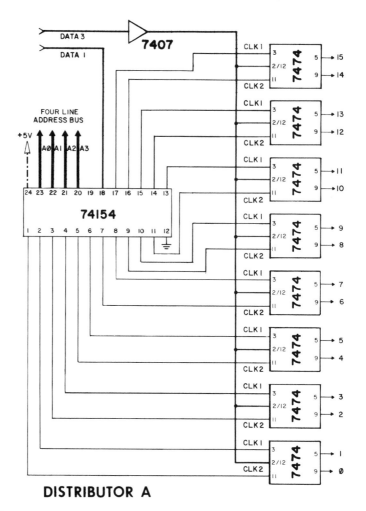

DISTRIBUTOR A

Figure 11.6 National Semiconductor 74154 data distributor. (Courtesy of LCDR H.R. Everett.)

sensitive and controlled by output PB7 of the 6522-2 VIA and is called latch Enable. When the Enable line is high, the latch contents reflects the value of the four-line address bus and subsequently holds that value when latch Enable goes low.

Drive and Steering Control

The wheel drive of the prototype robot was accomplished by two permanent-magnet gear motors which had separate forward and reverse windings. When the 12-V battery voltage was applied to the gear motors, their output drive shaft produced 12 rpm, producing a forward velocity of 16 ft/min. Since both drive motors are mounted on a single driving wheel, one on each side, the forward velocity mode requires one motor's forward winding and the opposite motor's reverse winding to be energized.

The prototype wheelbase is basically a boxlike cage constructed with aluminum angle. A dual-aluminum-angle inside frame supports a vertical steering column and thrust bearing in such a manner as to allow the steering motor to turn the driving wheel up to 80° left or right of center. Employing this method for steering allows the entire drive system to pivot around the steering column and to have a turning radius less than the overall length of the wheelbase cage.

The same type of gear motor used for driving the robot's wheel is used for steering control. Steering position is sensed by a belt-driven potentiometer mounted to the rear of the steering column. The sensing potentiometer is wired as a voltage divider across a regulated 5-V power supply to produce an output from 0 to 3.7 V as the wheel turns from right to left. This voltage is then fed to an isolating operational amplifier with its amplified output applied to another voltage-divider potentiometer set to apply 2.5 V (wheel full left) to pin 7 of a National Semiconductor ADC0804 analog-to-digital converter, as shown in Figure 11.7 and part of interface board 2.

The A/D converter was designed to accommodate an input voltage range of 0 to 5 V and a resolution of 8 bits in the free-running mode employed for robot control. In this application the A/D input voltage ranges from 0 to 2.5 V with 7 bits of resolution and the unused bit remaining low in normal operation. If the unused bit should ever go high, it is wired as an interrupt to signify that the system is no longer correctly calibrated and produces an A/D overflow. The prototype steering system was simplified to a point where only 4-bit A/D resolution was necessary to provide 16 discrete wheel position points throughout the total turning range. For this reason, the three least significant bits of the A/D converter were not used.

In operation, the 4 bits from the A/D converter are fed to the B inputs of a 7485 4-bit magnitude comparator and the desired steering position is fed to the A inputs directly from the 6522-2 PA∅ to PA3, as shown in Figure 11.8. The comparator compares the two numbers and indicates via its digital output if they are equal or else which is larger. The steering motor is directly controlled by these outputs and runs in the direction determined by which of the two outputs is greater. It continues to be run as long as the actual position of the steering wheel is not equal to the desired position.

As pointed out, this is a rather simple and crude steering system with no control over steering velocity. Nevertheless, it proved more than adequate for use in homing and navigational routines performed by the prototype. Considerable improvement could be obtained by coupling two 7484 comparators and picking up 3 additional bits of resolution from the comparators, for a total of 128 discrete positions of the steering wheel compared to the original 16 discrete positions. By monitoring the upper two most significant bits with exclusive-OR logic gates, the speed of the steering motor could be controlled by causing it to slow when the 2 upper bits are matched and to stop when all bits are matched (refer to Figure 11.9).

There is a considerable amount of frictional damping inherently present in this type of positioning system. It decreases the chances of overshoot, especially in the case of 16 discrete turning positions. However, as resolution is increased above the 4-bit control method, overshoot becomes a significant problem, and control of steering velocity becomes essential for stability. The most practical solution would be to use a small dedicated microcomputer to compare the desired position of the steering wheel to its actual position and control the steering motor's velocity with pulse-width modulation and continuously decreasing the steering motor's speed as a function of position error.

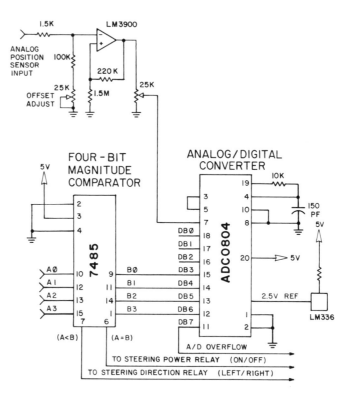

STEERING POSITION CONTROL SCHEMATIC

Figure 11.7 National Semiconductor ADCO804 steering control schematic. (Courtesy of LCDR H.R. Everett.)

It should now be obvious to the reader that control of any device requires more than just a computer with a parallel I/O and an RS-232 serial I/O. The presentation of the I/O expansion and steering circuits of Robart provides a vital knowledge base for any type of computer control function.

Robart's Head Control and Sensors

Robart's head, shown in Figure 11.9, differs from most robots in that it contains sensors for hearing, seeing, and obstacle avoidance. In addition, for maximum utilization of its head sensors, the head must be able to remain fixed at its center position or on command perform continuous movement by scanning back and forth over its maximum range of movement or track an external movement. A way to emphasize how the head and optical sensors function would be to describe how the robot determines its battery-charge status and seeks out its recharging station when its battery charge is low.

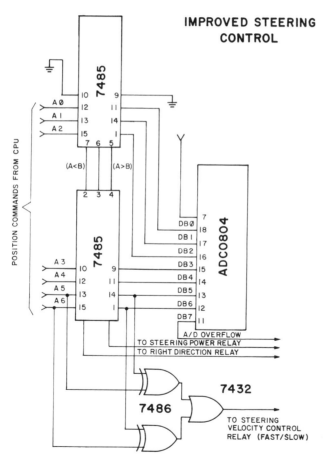

Figure 11.8 Improved steering control schematic. (Courtesy of LCDR H.R. Everett.)

Figure 11.9 Robart I's head
showing various sensors. (Courtesy
of LCDR H.R. Everett.)

Automatic Battery-Recharging Capability

For a robot to function in an environment hazardous to human beings, it must not only be immune to the hazards but must be able to support itself to a large degree. A prime consideration would be to be able to recharge its batteries when a low-voltage-battery condition is imminent and thus free the robot of the necessity of an umbilical cable to ensure adequate operating power. This requires an extremely reliable process by which the robot must locate its battery-recharging station and connect to its recharging system without being rendered ineffective by unforeseen obstacles or changes in its environment. At the same time, the circuit complexity must not take precedence over its primary functions.

The requirement for simplicity and reliability rules out the standard approach to the battery-recharging problem; trying to align a power plug on the robot with a mating receptacle on the charger. Although this can be accomplished with rather complicated hardware and software, it is by nature susceptible to many complications. The need is for a method of locating the charger and making electrical contact independent of the direction of approach without complicated alignment procedures and assures good electrical contact repeatedly.

Automatic Scan-and-Track System

A number of charger station location methods were evaluated to determine the most reliable method under adverse conditions. An optical tracking system offered the most positive identification method while maintaining simplicity in implementation. The charger must be equipped with a visual beacon that can be turned on and off by radio signals transmitted by the robot as a means of verifying the charger location beacon where other forms of lighting exist. A second method would employ a pulsed-tone-modulated infrared beacon on the charger which the robot detects by a tone decoder circuit. This eliminates the need for the verification step that would be necessary when an incandescent lamp is used as the charger beacon. Since the infrared light is invisible to the eye, it could be energized continuously and thus eliminate the need for a radio link between the robot and the battery-charging station. Once the robot has located the battery-charging station, by either the incandescent or infrared beacon methods, it must be capable of tracking the beacon while moving toward it.

The prototype model robot (Robart I) implemented tracking as an integral part of the hardware circuits located on the optical board and interface board 5, which control the position of the robot's head. The head-positioning motor is directly controlled by the optical board, and interface board 5 evaluates the CPU commands in conjunction with other inputs to determine the head control mode. This circuit (Figure 11.10) allows for three modes to control the motion or position of the head:

1. Scan mode
2. Track mode
3. Position mode

In the scan mode, the head sweeps back and forth over the 100° full right and full left of its centerline. This action is controlled by the scan flip-flop on the optical board mounted inside the robot's head. The scan flip-flop is set and reset by limit switches, at both extremities of head travel, causing the head motor to reverse direction and the head to scan in the opposite direction. This mode is selected by subroutine Scan, which sets the Scan Enable line high and the other two control lines low. When the Scan Enable line goes low and the Position Enable is also low, the head will be immobilized. All three lines are set low by subroutine Scanoff.

If the Position Enable line is high, the head will seek the position stored by the CPU in the 7475 4-bit latch on interface board 5. A 7485 4-bit comparator compares the command stored in the 7475 latch with the output of the head position AID converter and positions the head accordingly in the same way that similar circuits on interface board 2 position the drive wheel.

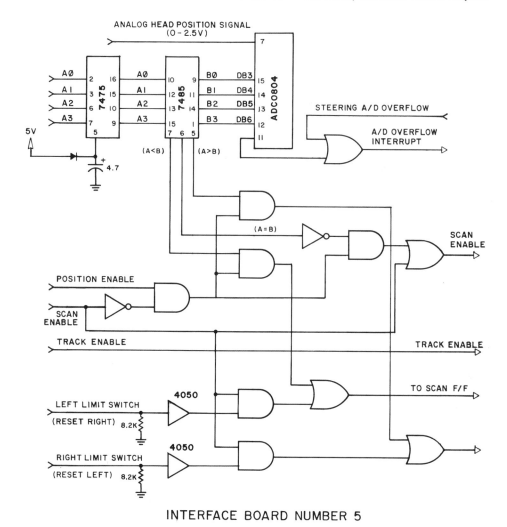

INTERFACE BOARD NUMBER 5

Figure 11.10 Interface board no. 5.

This head-positioning circuitry is automatically gated out whenever the Scan Enable line goes high. The latch is loaded from the interface four-line address bus and enabled by PB7 or PB2 on 6522-2. These operations are automatically performed by subroutine Latch, which takes the desired head position command from the Y register. In this manner, the head can be made to center itself after a scan operation or seek any of 16 fixed positions on command.

When Track and Scan Enables are high and the Position Enable low, the system functions in the tracking mode and the head looks for, locks onto, and follows the homing beacon on the battery-recharging station. It should be noted that the system will track any bright light source, and it is up to the software to ensure which light source is the homing beacon. The tracking sensor is composed of three photocells arranged in a horizontal array on the forward side of the head. Each photocell has a $\frac{1}{2}$-in.-OD 12-in.-long collimating pickup tube arranged with its center photocell tube parallel to the forward axis of the head and the outboard collimating tubes diverting 5° left or right from the forward axis of the head. Inasmuch as the head cannot tilt up or down, the photocell array must be mounted at the same height as the battery-recharging beacon (refer to Figure 11.9).

The optical board, shown in Figure 11.11, provides three digital outputs pertinent to the tracking process:

1. Target output
2. Point-source output
3. Range output

The optical board target output goes high of any of the three comparators associated with the photocell array goes high, while the point-source output reflects the status of the center photocell only. The range output indicates the relative distance from the homing beacon. The CPU communicates with the hardware associated with tracking the homing beacon with six lines, three control and three output.

The tracking process is initiated automatically by a low battery condition through alteration of the Behavior Selection procedure in such a way as to terminate the routine in process. When the battery condition goes low and remains below the set point for more than 5 s, a flip-flop on the monitor board changes state and triggers an interrupt. The IRQ routine, which handles the channel B interrupts, disables the low-battery interrupt and sets Register Next to set the docking routine. The on-board radio transmitter is activated to turn on the homing beacon on the battery-recharging station and the CPU enables the Automatic Scan and Tracking circuitry while dropping Position Enable line low. The head begins to scan left and right seeking a point source of light with sufficient intensity to trigger the photocell comparators. This action will continue as long as the Scan Enable and Track Enable are held high by the CPU and no light source is detected. If any of the three optical comparators goes high, indicating acquisition, the scan flip-flop is gated out automatically and the tracking inputs take over control of the head-positioning motor.

Tracking inputs to the optical board circuitry come from the left and right photocells in the array. If either photocell comparator indicates a greater light intensity than that of the center photocell comparator, the appropriate left or right motor winding will be energized and the head will turn toward the higher light level until the center photocell regains the higher-

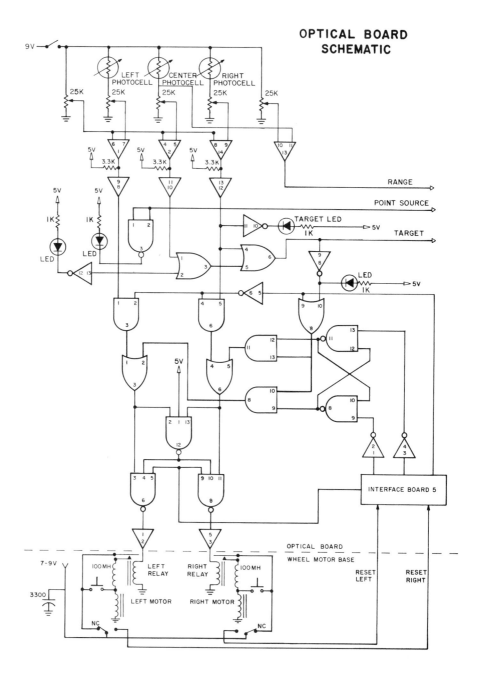

Figure 11.11 Optical board schematic.

intensity output. If by chance both left and right photocells show intensity greater than the center photocell, both inputs will be gated out and the head remains motionless. All of this happens only if at least one of the photocell outputs is above the adjustable set point provided by the background light bias circuitry on the optical board; otherwise, the system reverts to the scan mode and searches for a bright light source. Any comparator output signaling intensity above the set point gates out the automatic scan mode and the tracking mode takes over. When the photocell array indicates that the head is correctly positioned toward the light source, the scan motor windings will be deenergized.

The three collimating tubes on the photocells limit their field of view to relatively small regions. It is important that the homing beacon's height be positioned to center vertically within the detection zone of these tubes when they are pointed at the battery-recharging station. This results in a high signal-to-noise ratio as the head scans back and forth in search of the homing beacon.

The background light bias circuit provides the photocell comparators with a reference voltage above which the photocell output is probably due to a point source of sufficient intensity as to be the homing beacon. The original design provided a bias-adjust potentiometer to manually set the reference voltage. This method proved inadequate due to oversensitivity as the robot neared the beacon. If the bias was set low enough to allow detection of the beacon at long range, the system would saturate as the robot came closer to the beacon and all comparators would go high when the pickup tubes pointed in the direction of the beacon. Figure 11.12 shows an improved ver-

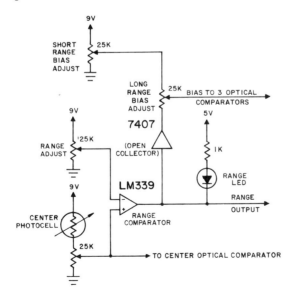

Figure 11.12 Photocell long range and short range sensitivity adjust. (Courtesy of LCDR H.R. Everett.)

sion of the optical sensor comparator bias circuit using two manually set reference points, one for long range and another for short range, that overcome the problem of saturation as the robot nears the battery-charging station. A fourth photocell comparator was added to monitor the center photocell output and automatically effects the bias changeover when its output signals that the robot is within 3 ft of the beacon. A range output line was added and is accessible by the CPU for use with other software routines to determine relative range (near or far) to the battery-charging station.

Once the CPU ascertains that the head has a lock-on (the Target voltage goes high), it must interrogate the source to verify that it is the desired beacon. Verification is accomplished by setting the Scan Enable line low and turning off the beacon via the on-board transmitter and observing if the Target output went low. An incorrect source will keep the Target output high. The Scan Enable is set low to keep the head from resuming its scanning if the source does not go out and the Position Enable is held low to prevent the head from seeking the position dictated by the contents of the 7475 latch when Scan is set low. If the source happens not to be the beacon, the CPU sends the Track Enable line low so that the source will be ignored, sends Scan Enable line high to reinitiate the scan, and waits as the head turns for the incorrect source to clear, signaled by the Target output dropping low. As soon as the Target output drops low, Track Enable is sent high again and a new source is sought. The process will be repeated until the correct source beacon is found, until three sources have been interrograted, or until a specified time limit has lapsed. The latter two indicate that the beacon is not in the immediate scan field (first and fourth quadrants relative to the centerline) and the robot will perform a U-turn to check the opposite direction.

Once the beacon has been located, its bearing and relative range are announced through speech synthesis and the robot begins to home in on the recharging station. The CPU repeatedly reads the head position and sends identical commands to the steering motor via interface board 2. As the robot turns, the head automatically tracks the source, and the relative bearing to the source decreases. The CPU subsequently reduces the turn angle until eventually the source is directly in front of the robot.

Docking System

Contact between the robot and the recharging station can be safely made if the voltages involved are in the range of the on-board battery and not the common 117 V of an ac power source. The lower-level voltages allow for contact surfaces to be exposed without presenting an electrical shock hazard. If the exposed contact surfaces are made axisymmetrical in respect to the vertical pole supporting the visual homing beacon, they will present the same target contacts on the approaching robot, regardless of the direction of approach. In this manner, the need for critical alignment to assure good electrical contact is eliminated.

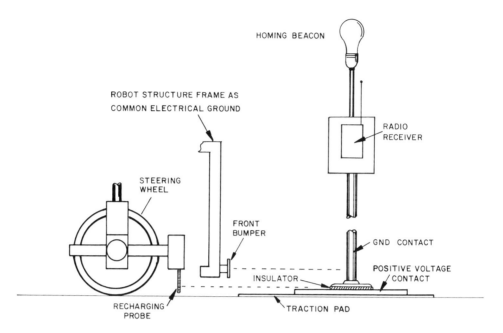

Figure 11.13   Battery recharging station and robot docking schematic.
(Courtesy of LCDR H.R. Everett.)

Figure 11.13 illustrates the axial positions of the two contacts to complete the recharge circuitry to the robot. A horizontal metal bumper attached to the front section of the aluminum-angle body frame serves as the common ground (negative) contact to the charger's vertical post common ground. One or more insulated spring-loaded recharging probes, attached immediately behind the front metal bumper to a bracket on the steering wheel, are used to make electrical contact to an insulated aluminum plate that serves as the charger's positive electrical contact. As the robot's bumper passes over the aluminum plate, moving toward the negative potential pole, the spring-loaded contact probes are brought into electrical contact with the positive potential aluminum plate and the robot's forward motion continues until the front bumper comes in contact with the vertical negative pole. As soon as the bumper contacts the pole, the electrical circuit is completed and recharge current flows into the battery. A relay is connected across the robot's recharge circuit which disconnects all power to the drive motors as soon as current begins to flow from the recharge station into the robot's battery. This serves as a backup for the software, which also deenergizes the drive motor when the recharge probes go high with respect to the electrical ground.

There are two power supplies associated with the recharging station. A relatively low-power 12-V source remains energized at all times, which supplies the receiver and decoder circuits to activate the beacon on demand from the robot. It also energizes the insulated aluminum plate through a resistor/capacitor network to a peak potential of 16.5 V and acts as a sensing voltage for the pickup probes to signal the microprocessor when the recharge circuit has been completed. As soon as the electrical contact has been made between the charger and the robot's battery, this voltage drops to 12 V. The drop in voltage is sensed by detection circuitry on the recharging station, which in turn activates the high charging power supply to furnish the high current required to recharge the battery. The high power supply automatically shuts off when the robot disconnects and the load is no longer sensed.

Robot Battery Monitor Circuit

The robot's battery voltage is monitored by an LM339 comparator, which sets a flip-flop, after a 5-s delay, when the voltage falls below an adjustable set point (see Figure 11.14). The time delay is to assure that the battery voltage was not momentarily pulled low by a stalled drive or steering motor. When the flip-flop changes states, an interrupt is generated which is read by IRQ channel B. The interrupt routine subsequently gates out the flip-flop and selects the docking routine.

The battery voltage is also monitored by a LM3914 dot display driver, which drives 10 LEDs to give a visual indication of the battery charge. The upper LED on this display also drives another comparator, which subsequently changes state when the battery is fully charged. This upper set point is also adjustable by the 50k potentiometer connected to pin 5 of the LM3914 dot display driver.

Collision Avoidance

The original prototype robot employed a six-level scheme of proximity and impact detection, implemented through numerous sensors installed at appropriate points on the chassis structure. An active near-infrared parabolic dish detector mounted on the head provided reliable detection of objects out to a range of 5 ft with a good bearing resolution of 2 in. of arc at maximum range. Additional long-range detection was provided by a forward-looking sonar and a 10-channel active near-infrared proximity detection system to provide close-in detection up to 18 in. Tactile sensors consisting of projecting feelers at critical points around the base periphery sensed impending collisions, and contact bumper sensors situated all around the base and body alerted the CPU to actual contact. The final backup monitored drive-motor current continuously for overload conditions.

The software dealing with all sensor data is divided into two basic groups, IRQ interrupt routines and the main control program. The main pro-

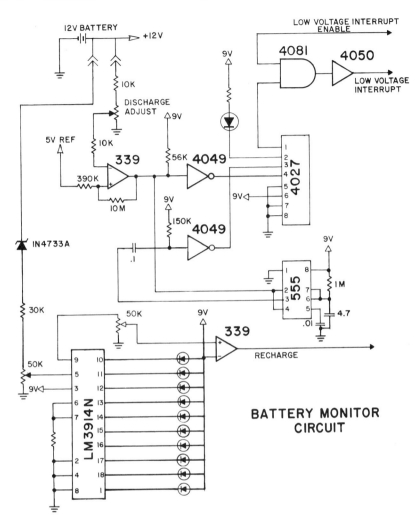

Figure 11.14 Battery monitor schematic. (Courtesy of LCDR H.R. Everett.)

gram handles the navigational control of the robot and it is here that the actual planning take place in the accomplishment or the desired task or goal. All information from program or sensors, as well as navigational subroutines, are written in loop form to facilitate repetitive polling, sonar activation, and delay timing, with appropriate exit requirements built into each loop. Loops can be cascaded as needed in the execution of complex navigational routines. Software associated with collision avoidance interrupt routines deals primarily with sensor data depicting the robot's immediate or close-in environment.

The six methods of detection can be broken down into three basic categories:

1. Ranging
2. Tactile
3. Internal

Category 1 inputs are read by software making up the navigational loop being executed, with resultant data used to alter course in a planned fashion. Examples of category 1 are sonar and near-infrared proximity detectors. However, these inputs are used to generate IRQ interrupts, to which the robot responds in a programmed reactionary fashion designed to clear the detected obstruction, based on the obstruction's location. These responses are implemented within the interrupt routine, upon completion of which control is passed back to the navigational loop, with the robot now clear of the detected obstruction.

Category 2 includes the feelers and contact bumpers, which also generate interrupts that move the robot away from an object with which it comes in contact.

Category 3 includes the drive-motor overload sensor, which serves as the last-resort detector, generating an interrupt command which reverses the drive motor in hopes of clearing the obstruction. On completion of the pre-programmed interrupt routines, the original drive-motor command is restored to the controlling circuitry and the robot proceeds along the programmed navigational path.

Sonar Collision Avoidance System

The original prototype collision avoidance system employed a National Semiconductor LM1812 monolithic sonar transceiver circuit operating at 21 kHz. The circuit, shown in Figure 11.15, provided range information to objects directly in the robot's path of travel, backed up only by the contact bumper system for impact detection. Initial tests showed that the LM1812 system was somewhat limited. Whereas large objects such as walls and furniture were reliably detected, smaller objects often passed unnoticed below the pattern of the sonar beam. Additionally, no information was provided by the sonar system as to preferable course alteration to avoid an obstruction in the navigational path. This was mainly due to the fixed position of the sonar's transponder, which did not permit scanning.

The first method to overcome the small-obstruction limitation of the long-range sonar sensor was to implement a number of feeler probes around the perimeter of the robot's base. These feelers extended 6 to 8 in. from the base and were configured to provide a normal highly TTL-compatible output which went low if deflection was sensed in any direction. The probes were flexible enough to prevent damage from contact to the robot or obstruction.

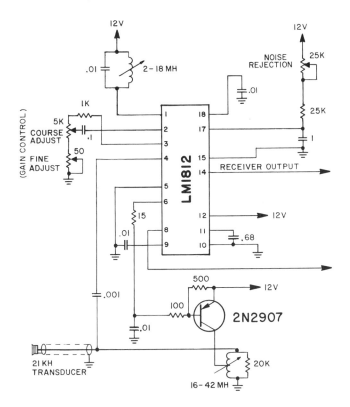

LMI8I2 SONAR SYSTEM

Figure 11.15   National Semiconductor LM1812 sonar schematic. (Courtesy of LCDR H.R. Everett.)

The intend in using the probes was to provide advance indication of an impact in time to alter course and avoid actual collision. Although the feeler probes were an improvement over the limited obstruction data from the sonar system, it remained a very crude approach to collision avoidance. One of the main problems with the feeler probes was the inertia of the feelers, creating false activation of the detection circuitry as they were jarred during motion of the robot.

The addition of four near-infrared proximity detectors provided obstacle information in front, or behind, and to each side of the robot within their limited areas of protection. Each IR sensor covered a cone-shaped region extending out approximately 12 in. The IR proximity sensors proved extremely effective in detecting objects within their field of view. The first improvements to the near-infrared proximity detection system was made by increasing the number of transmitter/receiver units from 4 to 10. Nine of the 10 units were installed in the first and fourth quadrants relative to the robot's centerline, for forward protection, because the majority of movement is in the forward direction. When collision avoidance routines call for reverse motion, the robot simply backs into space previously occupied, and the odds on having a rear impact are greatly reduced. Mounting the detectors in vertical rows resulted in an increased field of coverage and were simply hardwired to the same input in an OR configuration. Those that provided horizontal resolution of the obstacle location were kept separate from each other and read individually. The availability of this new sensor information to the CPU made possible more intelligent reactions to impending collisions.

Operation of the Near-Infrared Proximity System

The near-infrared proximity detection system consist of a centrally located driver/detector board with indicating LEDs and remotely mounted transmitter/receiver units, relocatable for optimum placement. The driver circuit

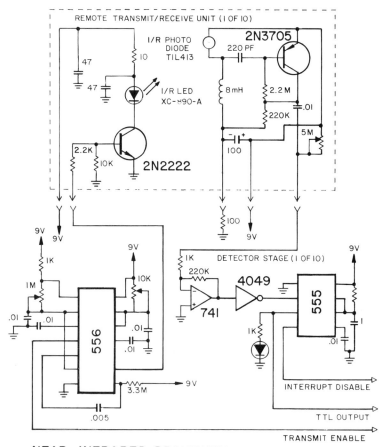

NEAR-INFRARED PROXIMITY DETECTOR

Figure 11.16   Near infrared proximity detector schematic. (Courtesy of LCDR H.R. Everett.)

is built around two identical pulse generators, each producing a square wave with a pulse repetition rate of 1.7 ms. The square waves then pulse the 2N2222 transistor into saturation, which gates the XC-880-A, a high-power gallium aluminum arsenide LED, to generate an infrared emission at a frequency of 880 nm. A 47-mF electrolytic capacitor and a 10-Ω current-limiting resistor are configured to supply current in excess of 2 A to the XC-880 LED for a very short pulse period and produces a high-IR-intensity beam in a narrow cone pattern. The two pulse generators that pulse the XC-880-A are alternately enabled by a 555 astable multivibrator which generates a 1-Hz Enable pulse. This results in reducing power consumption by a factor of 2 and eliminates pattern overlap where two adjacent XC-880-A LEDs are used. Figure 11.16 provides the schematic diagram for both the transmitter/receiver circuits and the main pulse generator section.

The receiver circuit consists of a TIL413 photodiode which incorporates a built-in filter and lens system to provide a cone-shaped detection field of approximately 45°. The output of the detector is amplified through an L/C differential network and fed to a 741 operational amplifier which subsequently produces a positive-going spike for each burst of returned infrared energy detected. The pulse output of the 741 is inverted by a 4049 and used to trigger a 555 monostable multivibrator. This acts as a pulse stretcher and produces an output pulse of approximately 100 ms and illuminates a monitoring LED. The 555's output generates an interrupt on IRQ channel A and is then read by data selector A. There are six receiver channels which are commonly enabled or disabled by distributor A, output 4.

ROBART II (1982–1983)

In any research project, the prototype development is a major learning experience. Original concepts and designs evolve into final concepts through modifications required to eliminate defects or create new approaches to the end performance from the trial-and-error effort in the prototype stage. This is evident when one compares the prototype Robart I to Robart II, which was developed more than a year later.

Robart II stands just 4 ft tall and measures 17 in. across its base. It employs a control hierarchy of six on-board 6502 microprocessors which control all sensor and mobility functions for navigation, collision avoidance, and environmental awareness. These include six ultrasonic rangefinders, 50 near-infrared proximity detectors, a long-range near-infrared range finder, and various sensors used to detect special alarm conditions, such as fire, smoke, toxic gas, flooding, vibration, and intrusion.

A frontal view of Robart II (Figure 11.17) shows five ultrasonic transducers on the body and one on the rotatable head. The long-range near-infrared rangefinder and its parabolic reflector is centered on the top of the head.

Three infrared true-motion detectors, two forward and one to the right side, can be seen just below the head. Figure 11.18 shows the body skin opened to exposed to the card cage, which houses the six computers and driver circuits. Figure 11.19 shows the removable control module with its computers mounted on the external side racks and the internal card racks fully loaded. At the bottom front is the on-board metering system for circuit testing.

The entire system reflects a superior design developed from the learning process in the Robart I prototype development stage. For example, compare the wheel-drive system of the prototype period with the Robart II wheelbase shown in Figure 11.20. Note that the steerable drive wheel has been replaced by two motor-driven 8-in. rubber-tire wheels with stability accomplished with the aid of swivel casters mounted front center and rear center of the wheelbase cage. This configuration permits the robot to spin about its vertical axis for marked improvement in mobility. The two wheel motors are controlled through pulse-width modulation and synchronized by high-resolution optical encoders attached to the motor armature shafts. The optical encodes

Figure 11.17   Front view of Robart II. (Courtesy of LCDR H.R. Everett.)

Figure 11.18 Robart II with skin doors open to show interior electronics. (Courtesy of LCDR H.R. Everett.)

Figure 11.19 Robart II's removable circuit card and computer cage. (Courtesy of LCDR H.R. Everett.)

Figure 11.20 Robart II improved wheelbase. (Courtesy of LCDR H.R. Everett.)

supply precise displacement and velocity information to the wheel controller computer for navigation and collision avoidance maneuvering. A low-level 6502-based controller handles all drive and steering functions on command from the top-level 6502 microprocessor.

Approximately 256 internal checkpoints are constantly monitored to check circuit performance, system configuration, operator-controlled switch functions, cable connections, distribution bus, voltages, and speech output generated by the self-diagnostics to advise of any difficulties. A 1200-baud serial on-board RF link can be used for telemetry or specific overriding of program commands from an observer at a remote terminal.

Considerable progress in industrial robots as well as robots such as Robart have been made at the university level since 1967, and it is anticipated that development projects of a similar nature will continue in the future as industry expands its use of computer-controlled machines, processes, and robots.

REFERENCES

Anderson, R. L. "Self Contained Independent Mobile Robot" (SCIMR) University of Pennsylvania, Moore School of Electrical Engineering, 1981–82.

Coles, S. L., Raphael, B., Duda, R. O., Rosen, C. A., Garvey, T. D., Yates, R. A., and Munson, J. H., *Application of Intelligent Automata to Reconnaissance*, Technical report, Stanford Research Institute, November 1969.

Everett, H. R., "A Computer Controlled Sentry Robot: A Homebuilt Project Report," *Robotics Age*, March/April 1982.

Everett, H. R., "A Microprocessor Controlled Autonomous Sentry Robot," Master's thesis, Naval Postgraduate School, October 1982.

Moravec, H. P., *Robot Rover Visual Navigation*, UMI Research Press, Ann Arbor, Mich., 1981.

Moravec, H. P., *The Stanfort Cart and CMU Rover*, Technical report, Robotics Institute, Carnegie-Mellon University, February 1983.

Nilsson, N. J., and Rosen, C. A., *Application of Intelligent Automata to Reconnaissance*, Technical report, Stanford Research Institute, February 1969.

Scanlon, L. J., *6502 Software Design*, 1980, pp. 180–218.

Zaks, R., *6502 Applications Book*, 1979, pp. 15-63.

CHAPTER REVIEW QUESTIONS

1. Did the robot Shakey have computer control, and if so, was it an on-board computer?

2. What form of the video image did Shakey use to navigate?

3. What does the acronym SCIMR represent?

4. Has sonar been a reliable navigating sensor to mobile robots? Explain.

5. What was the name of the first thesis robot to have multiple sensors?

6. What was the primary purpose of this robot?

7. What was the microprocessor used on the SYN-1 computer, and how many bits lines did it have in the address bus?

8. During the time Robart is under the control of a terminal, is it capable of movement?

9. What was the word count vocabulary in EPROM used with the Digitalker model DT1050?

10. Did the Digitalker EPROM require memory space in the SYM-1 computer?

11. How many 8-bit registers does a 6522 VIA contain?

12. How do the addresses of a VIA appear to the SYN-1 CPU?

13. What does the direction register do for VIAs or PIAs?

14. Several PIA/VIA registers were not discussed in detail. What functions do these registers perform?

15. In the prototype Robart I, what function required the most I/O lines?

16. On the parallel interface expansion board, how many data selector devices are used?

17. In the description of the data distributors (74154), reference is made to three control

lines, whereas the schematic shows that only two control lines were used. What is the purpose of the third control line?

18. How were the wheel motors reversed on Robart I?

19. How many discrete steering-wheel position points were obtainable with 4-bit A/D resolution?

20. What is the standard approach in connecting a battery to a battery charger in a robot?

21. What was the purpose of the radio transmitter on Robart I?

22. What navigational sensor circuits were included on Robart I's optical board?

23. What device reset Robart's head direction flip-flops when it was in the scan mode?

24. What was the purpose of installing tubes in front of the photodetectors in Robart I's head?

25. How long would the robot's head continue to scan once it received a Scan Enable signal?

26. What would happen of both left and right photocells show light-intensity outputs but the center photocell does not?

27. What improvement in photocell detection is obtained with collimating tubes in front of the photocells?

28. In the original prototype design, what happens when the photocell comparator bias is adjusted for long-range detection and the robot nears the homing beacon?

29. What advantage was obtained by the design of the robot's battery-recharging station?

30. What action signals the charging station to switch in the high-power charging power supply?

31. What is the purpose of the 5-s delay before enabling the low-battery routine?

32. How many computer CPUs are used in Robart II?

33. What major change was made in Robart II's wheelbase design over the prototype wheelbase?

GLOSSARY

Autonomous. The capability of the mobility to perform specific functions under its own power; functioning as a independent, self-controlled mobile machine.

Collimate. To bring into line or make parallel; to reduce a wide angle of view to a smaller angle of view by means of a mechanical device (e.g., reducing the wide-angle vision of the human eye by viewing through an optical device such as a telescope or binoculars).

Decrement. The process whereby the contents of a given computer register is decreased by one or more bits of data.

Enable. To make ready, energize, or start an electronic or logic device or circuit.

Hierarchical. Any system of things ranked one above another; one computer whose commands have priority over a number of dedicated function computers.

PIA. Peripheral interface adapter.

Scan. To move in a continuous manner over a fixed path or angle; a motion from left to right, reversing at the right limit, and returning right to the left limit and reversing again.

Software. A series of commands, instructions, or data, stored on punched cards or magnetic media, used to control a computer in performing a specific task.

Sonar. An acronym for *so*und *na*vigation and *r*anging: the use of audio tone pulses, above the range of human hearing, projected outward from a transponder and receiving back any echo of the tone reflected from an object in the path of the transmitted pulse.

Track. The action of following a moving object while keeping the object in the center of vision by rotating the view point; conversely, following a fixed object and keeping it in the center of the view point of a moving device, by rotation of the view point on the moving device.

Transponder. A mechanical or crystal device capable of changing an electrical signal into a mechanical signal or a mechanical signal to an electrical signal. For example, an audio speaker is actually an electromagnetic/mechanical transponder which is capable of changing an electrical signal into vibrations in the ambient atmosphere to produce sounds detectable by the human ear.

VIA. Versatile interface adapter.

12

Imagineering: The Future Applications for Computer Control and Robotics

Impossible is a word only found in the dictionary of fools.
Napoleon I

It would appear that the principal interest in the computer by the medical profession has been the discovery of cyberphobia, the fear of computers. This is surprising in view of the fact that computer stimulus of nerves and muscles has been used with considerable success to enable some paralyzed patients to regain limited mobility of their legs—to the extent of actually being able to ride a bicycle or walk. In cases where a patient has been totally paralyzed and has lost the ability to speak, a computer system can be coupled to computer vision to read a patient's eye movements and respond by vocalizing the patient's needs or call for assistance. We have the capability now to couple computer speech recognition to a computer-controlled wheel-drive and navigation system so that quadriplegics can verbally instruct their wheelchairs to move in any direction.

As a diagnostic tool, the computer, interfaced to computer vision, has proven its ability to diagnose medical X-rays as well if not better than a physician trained in X-ray diagnosis. Given X-rays diagnosed weeks or months previously, the computer arrived at the same diagnosis 85% of the time compared to the physicians' 50%. On the brighter side, the current emphasis on the use of biomedical electronic equipment may soon open the door for the computer as a trainable medical tool.

Computers have now been implemented into our public school systems as teaching aids at all levels, from kindergarten through high school, and short courses in computer awareness have become a requirement for graduation in a majority of our states. It would seem logical that the popular low-cost microcomputer could also be a valuable medical therapeutic treatment for quadriplegics. The voice-entry technology makes personal computing accessible to the physically disabled and enables them to write letters, make telephone calls, use electronic mail, access information networks and bulletins, program or operate a computer, and even control their environment by being able to turn on or off room lights or TV, adjust heating or cooling thermostats, and even control the movement of a wheelchair. The disabled person could have a computer permanently installed on a wheelchair so that the chair could be accessed anywhere in the range of the patient's voice.

VOICE-ENTRY TERMINAL

All these possibilities can be accomplished through a little *imagineering.* Starting with the Scott Instruments product called the Shadow/Vet voice-entry terminal, a disabled person would create his or her own speech control vocabulary by teaching the Shadow/Vet terminal to recognize his or her speech patterns. A vocabulary consists of 40 words which may be a phrase, a command, or an alphanumerical character or digit. Vocabularies of as many as 40 word groups can be stored on a disk for instant recall. The voice recognition system is taught the user's voice characteristics regardless of language, dialect, or utterances resulting from speech impediments.

Shadow/Vet is compatible with most computers using BASIC, Applesoft, Pascal, FORTRAN, or machine code. The basic requirements include a minimum of 48k RAM, one disk drive, one peripheral interface slot, and DOS 3.3 for system initialization. Shadow/Vet hardware consists of a microphone headset and a speech amplifier unit, shown in Figure 12.1. The speech

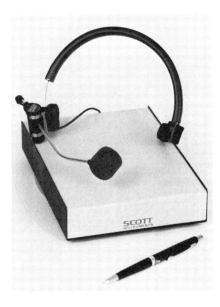

Figure 12.1   Shadow/Vet speech amplifier and microphone headset for hands-off voice control. (Courtesy of Scott Instruments.)

amplifier's output connects to the audio input of the control card installed in the computer's interface slot. The Shadow/Vet's hardware and software is applicable to Apple II, Apple II Plus, Radio Shack TRS-80 models I and III, or any computer with compatibility with Apple or TRS-80 computers.

COMPUTERIZED WHEELCHAIR DESIGN PROJECT

To better illustrate the term *imagineering*, suppose that we borrow a few ideas from Chapter 11 and combine them with the Shadow/Vet voice terminal to make a practical application of voice control of a vital medical application: quadriplegic patients' wheelchairs. Figure 12.2 illustrates how a wheelchair could be adapted to incorporate computer-controlled mobility and navigational functions under the control of a voice terminal.

The first modification would be to install a platform as support for wheel-drive motors and a rechargeable gel 12-V battery. The platform bridges the wheelchair's lower siderails between the two vertical posts sup-

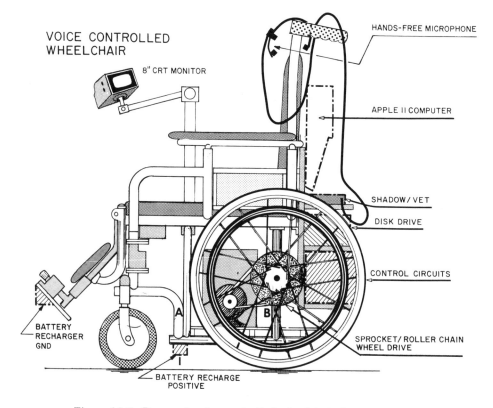

Figure 12.2 Proposed voice controlled wheelchair design project.

porting the armrest and the rear wheels. The platform must be rigid enough to support the added weight of two wheel-driving motors and a 12-V battery. The platform may be cut from aluminum plate stock ranging from 0.125 to 0.250 in. thick and cut to fit snugly between the upright posts, A and B. Depending on the amount of weight to be supported, the 0.125-in. platform aluminum may require bracing with 0.500-in. aluminum angle. After fitting the motor support plate across the wheelchair's lower rails, the plate is then secured with two 12-32 × 2 in. machine screws in each lower rail.

The wheel-drive motors presented in prior chapters were connected directly to the driven wheel. The weight and power factor for wheelchair wheel-drive motors requires a positive traction coupling by sprocket gears and roller chains.

Wheelchair Sprocket Chain Drive

The rear, hand-operated wheels are removed to attach a 10-in.-diameter 120-tooth sprocket gear, which is to be attached on the chair side of each wheel. To provide working clearance of the roller chain, between the wheel sprocket gear and the wheel spokes, it will be necessary to attach an 8-in. diameter 0.250-in.-thick aluminum spacer to the 10-in. sprocket gear on its wheel side. Should the wheel axle bolt be larger than 0.50 in., the sprocket gear's center axle hole must be drilled to the diameter of the axle bolt. In any case, the 8-in. spacer must be drilled to fit the bolt diameter. The wheel spoke hub contains bearings and cannot be used to attach the sprocket gear and spacer plate to the wheel. An alternative method can be used whereby the gear spacer plate can be drilled and tapped to attach metal clamps to lock the sprocket gear to the wheel spokes. It will be necessary to add a washer of similar thickness between the gear and the wheel axle support column to provide clearance for the roller chain and the axle support column. The accumulation of spacer and sprocket gear thicknesses will require replacement of the axle bolt with one approximately 0.750 in. longer.

Once the wheels have been equipped with the sprocket gears, two reversible 12-V dc motors can be mounted and a driver sprocket gear installed on its armature shaft. The selection of the motor sprocket gear will depend on two factors: the shaft diameter and the speed reduction desired. The ratio between the number of sprocket teeth of the driving gear and the number of sprocket teeth in the driven gear determines the speed-reduction ratio of the gear combination. The relation between driver gear teeth and motor armature shaft diameters is shown in Table 12.1.

With a driven gear of 120 sprocket teeth, the range of driver gear sprocket teeth in Table 12.1 permits reduction ratios as high as 12:1 to as low as 3:1 (120/10 = 12:1). However, the motor armature shaft diameter is a gear selection limitation as to maximum reduction ratios that can be used.

TABLE 12.1   Sprocket Gear Bore to Number of Teeth

Gear bore diameter (in.)	Number of gear teeth	Gear OD (in.)
0.250	8	0.176
	9	0.842
	10	0.917
0.325	11–30	1.075–2.504
0.500	31–40	2.583–3.295

After installing the dc motors and their sprocket gears, the motor gear is then connected to the wheel gears via a No. 25 standard roller chain. Sources for sprocket gears, roller chains, and design data are given in Appendix E.

Referring again to Figure 12.2, an aluminum angle framework is attached to the rear side of the wheelchair seat to form a shelf or cage support for the voice terminal, computer, and other control circuit boards. It would also be appropriate to include a modified version of the Robart I battery-recharging system. Figure 12.2 shows where the ground and 12-V positive contacts should be located for battery-recharging electrical contacts.

Wheelchair Sensors

If the computer-controlled wheelchair were a live project, it would be possible to present detailed drawings with size and layouts for metal plates, spacers, angle frame members, and so on. However, this is an *imagineering* project, where details of this nature are left to you to work out. Nevertheless, the project has validity and could be designed by a few top students using this section of the chapter as a guide. Thus, as an aid to design planning, instead of diagrams and schematics for the prototype computer-controlled wheelchair, the requirements to make the project become a feasible product are listed.

1. *Navigation*: to move as instructed on voice command, at a reasonable speed, with automatic doorway centering and collision-avoidance capability

2. *Utility function communications*: to operate in the voice terminal mode for computer operations in word processing and printing for correspondence, modem interfacing for network or bulletins, or access to local and long-distance telephone facilities

3. *Utility function entertainment*: to access, via voice command, computer educational or popular computer games software, control television receiver ON/OFF status, channel selection, sound/color/hue adjust, and antenna rotation for best picture when an external antenna is used instead of cable distribution.

4. *Utility function environment control*: on voice command, activate specific room lights ON or OFF, adjust thermostat for heat or cooling to the voiced temperature setting, detect and alert on the detection of smoke, fire, flooding, severe storms, or intrusion; automatically turn on all indoor and outdoor lights if intrusion is detected after dark.

5. *Utility recharge function*: to seek out recharge station on voice command or alert operator if battery status dictates recharge state by synthesizing the words "time to recharge battery"; should an intruder or hazards of smoke, fire, or gas be detected during nighttime recharging cycle, automatically disengage recharge cycle and navigate to operator's bedside to evacuate operator from building

There are logical approaches by which any or all of these utility functions can be carried out. Some of the external device control functions could be performed by a radio link between the wheelchair and the controlled device. Only one function would require the wheelchair to have some form of manipulator, and that is getting a quadriplegic out of bed to evacuate the premises in case of fire. In this case the patient must be wearing a harness which has an extending strap, terminated with a steel ring to which the manipulator could attach a power winch cable. In this manner the patient could be pulled around to a position where he or she could be slid or pulled backwards onto the wheelchair and to the normal seated position.

The guidelines for any project development are simple to follow:

1. *A market must exist for the final product.* In this case, a major market potential exists.

2. *The end product must be functional, comparatively low in cost, and virtually maintenance free.* The wheelchair would be functional, virtually maintenance free once all bugs had been corrected, and comparative in price to an average-price automobile.

VETERANS' ADMINISTRATION'S QUADRIPLEGIC WHEELCHAIR WITH MANIPULATOR*

One of the first steps in any project design or development is to study the state of the art existing prior to the start of the project. By a coincidence, the Veterans' Administration, Rehabilitation Development Services, in New York City has been developing a wheelchair manipulator for quadriplegics

*This section is based on [Mason, 79].

that should provide helpful background information to the computer-controlled wheelchair project.

In an effort to restore human dignity to totally disabled quadriplegics and to generate self-confidence, independence, and the ability to operate like any other human being, the Veterans' Administration Prosthetics Center, Rehabilitation Development Service, has developed a medical manipulator for quadriplegics in an effort to replace their lost functions of mobility and manipulation. The approach has been to solve the problem of patient loss by amplifying remaining capabilities. The quadriplegic's brain could serve as a functional control center if we could produce machines with which quadriplegics can effectively communicate. The manipulator pictured on a wheelchair in Figure 12.3 may provide the means to perform some of the activities

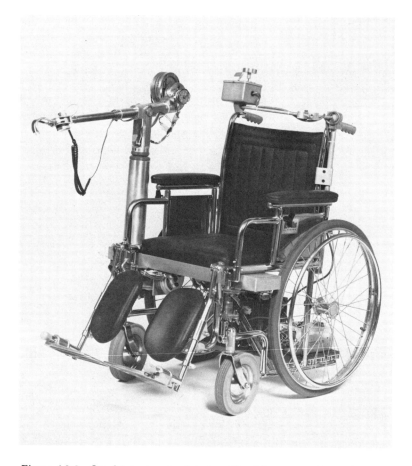

Figure 12.3 Quadriplegic wheelchair with chin operated manipulator control. (Courtesy of the Veterans Administration.)

for daily living. From the individual's point of view, any form of lost function replacement must appear *logical* and should appear *simple, repeatable, proportional*, and *reliable*. Function-loss-replacement devices may be *aided, trainable*, or *fully automatic* to reduce the user's input. The quadriplegic can provide control inputs to function-loss machines in a number of ways:

1. *Overt motion*: chin motion or head motion
2. *Overt force*: chin or head generated
3. *Electromyograms*: overt muscle activity of the forehead
4. *Voice*: spoken words or utterances
5. *Ocular*: control eye motion or position
6. *Electroencephalograms*: brain and thought waves

All these control sources have limitations, advantages, and disadvantages. Perhaps the ultimate control source may be obtained by a combination of two or more of the available sources for the quadriplegic's control center.

Development cost of the computerized wheelchair and manipulator in Figure 12.3 ran approximately $100,000 and production models are expected to cost about $10,000. Although the prototype model is approxi-

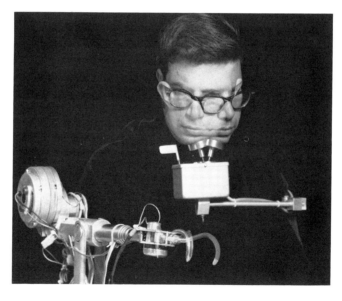

Figure 12.4   Chin movements to open or close the manipulator's fingers. (Courtesy of the Veterans Administration.)

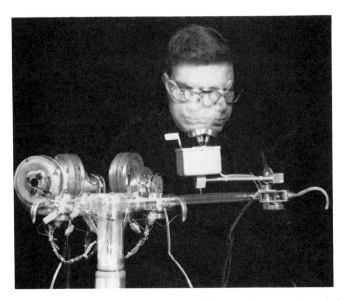

Figure 12.5 Chin movements to swing the manipulator from left to center and center to right. (Courtesy of the Veterans Administration.)

Figure 12.6 Chin movements to rotate the manipulator's fingers 180 degrees. (Courtesy of the Veterans Administration.)

mately 97% error free, there are openings for improvement. The voice terminal computer has only a 35-word vocabulary and the manipulator required chin pressure and movement for control. As an example of chin control of the manipulator, in Figure 12.4 the developer of the system, Carl P. Mason, demonstrates, by double-exposure photograph, the chin motion on the gimbal-style control to open and close the manipulator's gripper or fingers. Figure 12.5 is another multiexposure photograph that moves the manipulator left to right by duplicating the gripper chin motions but at a different chin pressure. Producing a rotating action on the gripper simply requires an up–down motion like nodding yes, as shown in Figure 12.6. The limits of the horizontal travel of the manipulator can be seen as the streak of light produced by a small lamp held in the gripper (Figure 12.7); its vertical motion limits are shown in Figure 12.8. Figure 12.9 demonstrates the reach or extension capability of the manipulator.

The mechanical development of a project like this is the "easy" part,

Figure 12.7 Limits of the manipulator's horizontal movements. (Courtesy of the Veterans Administration.)

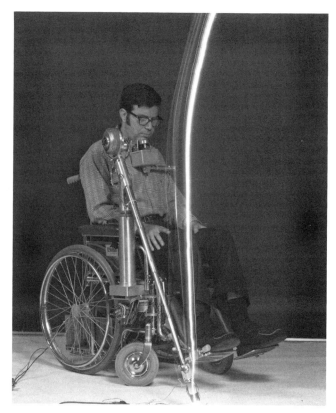

Figure 12.8 Range of the manipulator's vertical movements. (Courtesy of the Veterans Administration.)

but prone to be the most expensive, especially when each metal part must be virtually a hand-made fit through trial and error. For example, the motor that extends the manipulator obviously drives a flexible strip of metal, not unlike a metal measuring tape, that pushes out or pulls back the telescoping sections of metal tubes in the same manner that the disappearing automobile telescoping radio antenna raises and lowers. From the multiexposure photographs it is apparent that the roll axis and the finger action take place at the end effector's junction to the manipulator and is not produced by rolling the manipulator's telescoping tubes.

Imagine the self-satisfaction and pride of accomplishment for a quadriplegic fortunate enough to have such a computerized wheelchair and manipulator. After years of dependence on others, he or she could invite a visitor to play a game of chess, as shown in Figure 12.10.

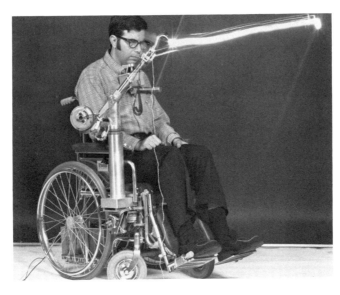

Figure 12.9 Angle of extension possible with the manipulator. (Courtesy of the Veterans Administration.)

Figure 12.10 Manipulator's developer showing how a quadriplegic could play chess with the manipulator equipped wheelchair. (Courtesy of the Veterans Administration.)

NUCLEAR POWER PLANT DAMAGE CONTROL ROBOT*

To the majority of the population, the publicity of the near meltdown at the Three Mile Island nuclear power plant and the present disposal of deactivated nuclear power plant dismantled highly radioactive materials (which may remain radioactive for 1000 years or more), poses the questions: What do we do when there is an accident in a nuclear power plant and the situation becomes critical? How do we safely store highly radioactive dismanteled nuclear power plants for a thousand years or more? How do we repair leaking radioactive coolant systems when the area has become so contaminated with radiation that human beings cannot enter to effect repairs? Not only have we had the "impossible" near disaster at Three Mile Island, we now have learned that a number of nuclear power plants have had spillage of radioactive materials and far too many reports of faulty inspection practices, faulty workmanship, and faulty construction materials found at nuclear power plants under construction. There must be some corrective action to safeguard the population instead of the usual cover-up and denials. The truth is that no provisions are included in any nuclear power plant design to safely allow human corrective action in highly radioactive areas. According to the Department of Energy, repair robots will not be required by nuclear power plants until the development of second-generation nuclear power plants some 25 years hence.

Damage repair robots are the only solution for entering highly radioactive areas to effect immediate repairs. However, we must know what has happened and where it happened before we can begin to think of what can be done to correct the situation. Of course, once a robot enters the contaminated area, it too becomes contaminated and cannot return outside for repair materials even if it were possible for the robot to make the repairs.

We have the technology now to develop the type of robot that could make nuclear power plant repairs in highly radioactive surroundings, even in totally flooded reactor rooms. However, to be workable it would be necessary for the repair materials and robot to be stored in critical areas before an accident happens. To be able to function in a radiation-contaminated environment, a repair robot must be constructed to resist the effects of radiation, heat, and submersion. It must be able to view the damage area under all adverse conditions and to transmit this knowledge to the damage control engineer so that it can be determined what repairs can be accomplished. It must have the mobility and manipulation to perform the work of repair even if the reactor area has been totally flooded. By its very nature, it would be extremely heavy, on the order of 20 tons (40,000 lb).

Figure 12.11 illustrates the author's model of a damage control robot

*This section is based on [Gupton, 80a, 80b, 83].

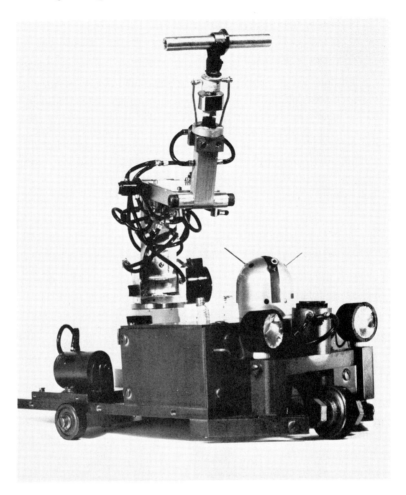

Figure 12.11   A model of a nuclear power plant damage control repair robot (DCRR). (Courtesy of Dilithium Press and *Robotics Age.*)

(DCR) or damage control repair robot (DCRR), which demonstrates some of the features a nuclear damage control repair robot should have:

1. The mobility wheels are driven by independent hydraulic motors. The wheels are solid metal with solid insulating tires. Wheel bearings are solid noncorroding metal with no hydrocarbon-based lubricant used.

2. The equipment support base must be structurally strong enough to support two industrial robots similar to the Cincinnati Milacron or Unimation heavy-duty robots and a number of interchangeable tool

hands. Independently steerable floodlights are mounted on the forward end of the mobile base to illuminate the damage area for video observations and to direct repair activities of the robot.

3. The actual robots can be standard industrial robots modified for heavy payloads, extra extension of their vertical reach, and an added modification to permit self-exchange of its work tool hands.

4. Control is by combination computer control of the robot work functions, aided by visual supervision from the damage control engineer. The control link between the damage control center and the robots is via fiber optic umbilical cables which also serve as a video observation link. Electric and hydraulic power to operate the robot are direction cable connections to plant systems. A self-retracting reel system prevents the robot vehicle from entanglement while providing maximum freedom of mobility within the reactor area.

5. On-board storage of power tool work hands or robot end effectors enables the robot to exchange one work tool for another as needed for metal plate, piping, or mechanical part replacement. Included would be an inpact or power wrench, a metal saw, an impact hammer, and arc welding, drilling, or nonwelding sealing. Figure 12.12 illustrates the work tool storage compartment with the manipulator in the act of changing a work tool.

Figure 12.12 The DCRR robot's work hand storage. (Courtesy of Dilithium Press and *Robotics Age.*)

The key to a nuclear damage repair robot will be the ability for it to change its own work tools, since its location in a contaminated area prevents human change of the work tool as required for present industrial robots. Figure 12.13 shows a practical bayonet/scabbard method for enabling a robot to change work tools. The bayonet connector is attached to the end of the manipulator and the scabbard receiver is attached to the work tool. In operation, the robot would index the manipulator to line up to a position where the current work hand would be stored and insert the tool into the storage compartment. The computer would then transmit a lock-pin release signal followed by a 90° counterclockwise rotation signal to the end effector. The lock pin is withdrawn from the scabbard by a solenoid to permit the bayonet to rotate 90° to disengage its locking flanges and withdraw. The computer next indexes the manipulator to a position whereby the now-empty bayonet can be inserted and locked to the next work tool required. Figure 12.14

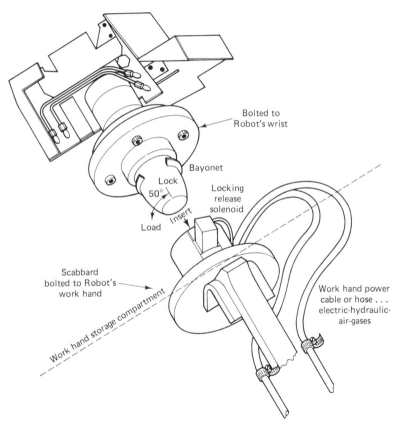

Figure 12.13 Bayonet/scabbard principle by which the DCRR robot could change its work hand. (Courtesy of Dilithium Press and *Robotics Age*.)

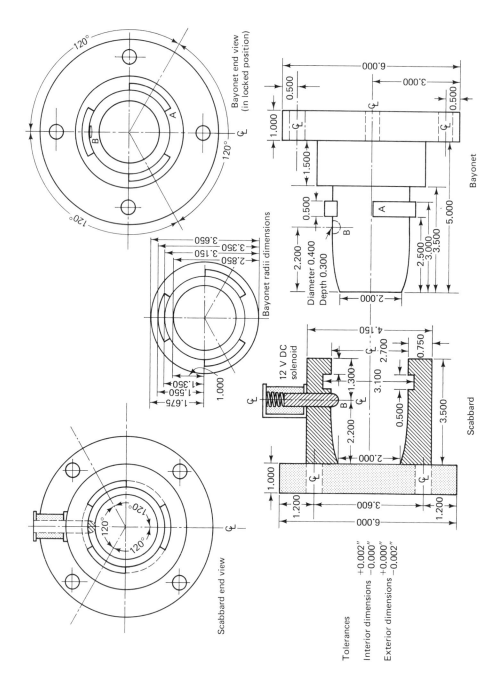

Figure 12.14 Bayonet and scabbard contours and dimensions. (Courtesy of Dilithium Press and *Robotics Age.*)

306

shows a diagram with dimensions to illustrate the size and details necessary to produce a bayonet/scabbard method for a robot to change its work tool hand.

MINE RESCUE ROBOT

Nuclear power plant damage control and repair is not the only immediate application for damage, accident, or disaster robot use. All too frequently we hear of mine accidents trapping workers deep with the bowels of the earth. Poison gases, fires, and collapsing tunnels take hundreds of lives each year. In many cases, a special "tunneling" robot could be used to reach trapped workers, bringing fresh air and opening an avenue of escape. Figure 12.15 illustrates an approach to a mine rescue robot. The robot's tunneling ability is created by a circular high-speed cutter wheel located at the nose of the rescue robot. The cutter tool is faced with diamond or Carboloy teeth, giving the robot the capability of cutting a 26-in.-diameter tunnel through stone, clay, or coal.

Spaced around the circumference of the robot are a number of water nozzles used to cool the cutting face and to flush all cutting debris clear of the tunnel. The robot is powered by eight wheels, mounted as tandem driving pairs every 90° around the robot body. To stabilize the robot in its self-cut tunnel, the top and bottom tandem wheels are staggered to the rear of the side wheels. Navigation of the tunneling robot is accomplished by increasing the wheel motor speed of the tandem-driven wheels on the opposite side of the robot to the direction of course correction. As an example, increasing the motor speed of the top tandem wheels would cause the robot's nose to be pushed downward, causing the tunnel cutting direction to inch downward slowly.

Hose and cable attachments at the rear of the robot provide power and control of the robot, water for cooling and flushing, and air or oxygen for the trapped miners once the robot tunnels through to them. The diameter of the rescue tunnel would be large enough to serve as an avenue of escape for the trapped miners.

SHIPBOARD DAMAGE CONTROL ROBOT

Perhaps the ultimate challenge for robot application would be to design a robot to repair ship hull ruptures resulting from a collision with another vessel, an iceberg, or an underwater barrier, or from enemy fire during wartime. The limited number of accidents of this nature dictates that a seagoing damage repair robot should have the versatility to handle the additional duties of a firefighter robot.

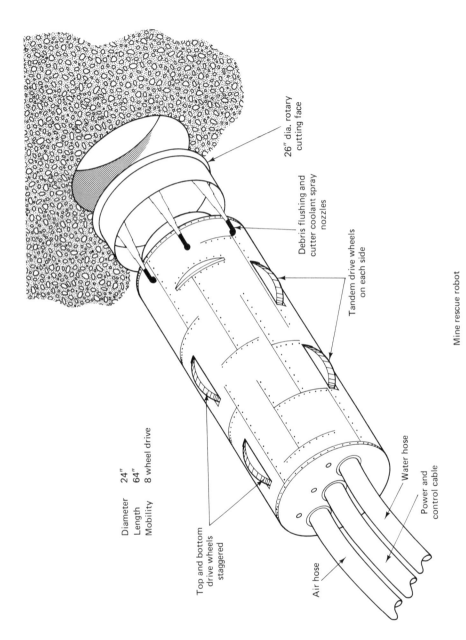

26" dia. rotary cutting face

Debris flushing and cutter coolant spray nozzles

Tandem drive wheels on each side

Top and bottom drive wheels staggered

Air hose

Water hose

Power and control cable

Diameter 24"
Length 64"
Mobility 8 wheel drive

Mine rescue robot

Figure 12.15 Mine rescue robot.

The design effort of this type of robot immediately encounters the major problem of how to maintain mobility when the damaged vessel has developed a severe list to one side. Mobility can be further impeded by wet or flooded metal decks and failure of the robot's wheels to secure traction. Imagine the robot mobility problem in a spaceship such as the *Columbia* Space Shuttle, where those on board encounter weightlessness and loss of vertical and horizontal orientation during the time the spaceship is in orbit. The end result is identical for a robot whether dealing with the listing of a vessel or weightlessness in space—its mobility is lost by the failure of its driving mechanism to gain usable traction. Two solutions immediately come to mind: return the listing ship to normal position by flooding compartments on the opposite side of the list; and when in orbit, create artificial gravity aboard the spaceship.

In the case of a list, the list is caused by the ship or vessel taking on seawater at a faster rate than the discharge from the ship's pump. If the list is severe, additional flooding to upright the vessel might be sufficient to sink it. Creating artificial gravity aboard a spacecraft requires the expenditure of limited fuel supplies to produce a gravity effect by having the spaceship rotate at a fixed rate, which would be halted prior to reentry. Using precious fuel in this manner could prevent normal mission course adjustments, including the critical reentry adjustments, and might cause the loss of the spaceship.

The solution does not lie in correcting the listing of the ship or the weightlessness in orbit but rather in finding a way for the robot to retain its traction under these conditions. Oddly enough, the way to improve robot mobility traction, as well as automobile tire traction, does not originate in a sophisticated scientific theorem but comes from a tiny tropical insect. This particular insect is a little smaller than the familiar ladybug, and lives on the palmlike leaves of a tropical plant. This insect has a hard shell, similar in appearance to a miniature version of a walnut shell. In motion, the insect appears to be a small shell with eight legs and feet encased in fluffy bedroom slippers. When endangered by attacking ants, the insect draws its legs under the armor of its shell and holds onto a leaf. Strangely enough, the powerful ant finds it impossible to dislodge this tiny creature from its hold on the plant's leaf.

Examining the insect's feet under a microscope reveals that the feet contain several thousand suction pads on hairlike appendages. When dry, the suction pads offer little adhesion to the surface, but if sprayed with oil from glands around the suction cup appendages, the suction power is considerable. Applying the principle of the insect's feet to robots would provide a method for improving traction on wet metal decks of ships or mobility in the zero gravity of space.

Figure 12.16 illustrates a robot designed for dual duty in shipboard firefighting or collision damage control. The first oddity is that the robot employs tanklike rubber-tread belts instead of the customary wheels. It ap-

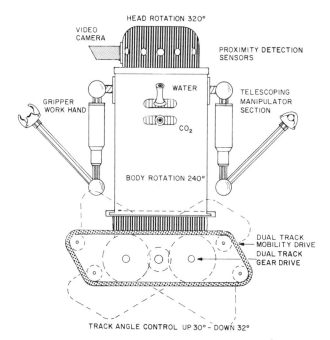

HEAD ROTATION 320°

VIDEO
CAMERA

PROXIMITY DETECTION
SENSORS

WATER

GRIPPER
WORK HAND

TELESCOPING
MANIPULATOR
SECTION

CO_2

BODY ROTATION 240°

DUAL TRACK
MOBILITY DRIVE
DUAL TRACK
GEAR DRIVE

TRACK ANGLE CONTROL UP 30° - DOWN 32°

Figure 12.16 Proposed fire
fighting and underwater repair
robot for shipboard disasters.

pears that its manipulators are located on the front and rear sides of its
body. For navigation through narrow passageways, the robot's body rotates
90° to reduce the width of the body and side manipulators. For firefighting,
the robot is equipped with water, CO_2, or foam for any kind of on-board
fire. The robot's head contains a video camera, floodlights, and proximity
detectors to be able to see and report to damage control the extent of fire
or collision damage. The proximity sensors aid the robot in navigation
through the ship's passageways. Head rotation is desirable in order to scan
the entire damage area for damage control evaluation.

Another odd feature shown in Figure 12.16 is the ability to tilt the dual
tracks up or down. The reason for this is that most Navy ships have odd-
shaped doors or hatches where the bottom of the hatch is at least 12 in.
above the deck. To make it even more interesting, the hatches are locked
(dogged) by a number of wedge-type levers. Encountering this type of hatch,
the robot can open the dogged locks with its manipulator and the front end
of the dual tracks can be moved up the bulkhead (wall) aided by a motor
geared incline control to raise the dual tracks. Once it begins to pass through
the hatch, the dual tracks then rotate downward to meet the deck on the
other side of the hatch.

To go from one deck to another, the robot discovers that ships have
gangways (stairs) going up or down between decks. The gangways are very
narrow and steep, to the point that they constitute virtually a free-fall obstacle to any robot design. The damage control robot merely locks its grip-

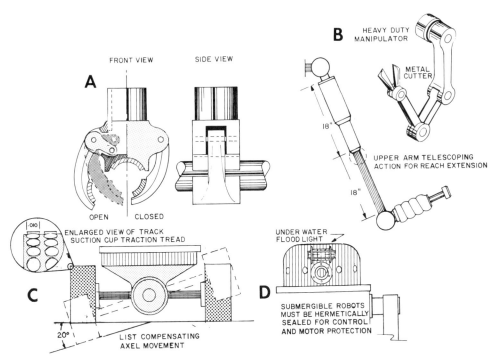

Figure 12.17   Naval robot's key features.

pers to the gangway's handrails and lowers its dual-tread tracks, braking its forward speed by increasing the gripping pressure on the handrails. The dual-tread tracks not only have up or down mobility, but also have ability to tilt, raising one track while lowering the opposite track. In this manner, the robot can remain in a vertical orientation even when a ship has a list of as much as 20°.

Figure 12.17C illustrates the tilt axle function as well as the suction-cup traction belt design and other functions related to the robot design in Figure 12.16. Figure 12.17A shows the design and friction pads of the gripper and how it clamps to the handrail to serve as brakes when going down a gangway. Figure 12.17B shows the telescoping capability of the manipulator to extend its reach capacity, and Figure 12.17D points out the waterproof floodlight and hermetical sealing requirements for working submerged in water.

It is very important that damage control or firefighting robots have the capability of relaying television images back to damage control center, so that the control personnel can determine how to control the fire or effect repairs to the hull and control the robot accordingly. Under emergency conditions of fire and hull damage, fully automated robot control would not be practical and must be supplemented with visual input and operator control via an umbilical cable. In view of the need for the robot to provide water, CO_2, or foam in fighting shipboard fires, these too must be supplied to the robot through connecting hoses. The communications cable and fire fluid hoses could be on retracting reels attached to the robot and payed out as required to reach the fire site. When the fire is put out, the robot merely rewinds the hose onto the payout reels as it returns through the ship's passageways.

SECOND-GENERATION WORK ROBOTS

It is time to design and construct a second-generation industrial robot to take over the hazardous jobs outside the factories of the manufacturing community. Every day wasted in developing work robots with the capability of performing specific functions in high places means more human injuries and loss of life. Some of the present hazardous jobs performed by human beings but capable of being done by robots include:

1. Washing windows in tall buildings
2. Painting tall structures (e.g., bridges, towers, steeples, smokestacks, water storage tower tanks)
3. Fire rescue at airports, oil wells, buildings, and so on
4. Underwater searches for sunken vessels or drowned persons
5. Underwater mining, coal mining, and mineral deep mining
6. Space repair of satellites

Future computer-controlled applications include automatic guidance automobiles, climatic control for homes, offices, and industries, and a marriage of computer-controlled machines and robots for full automation of manufacturing. The latter is virtually an accomplished fact in foreign production of fabrics and steel. We already have on the drawing board designs for a computerized/robotized aircraft factory that needs only one human being to switch the power on or off to start or stop the assembly lines. Even more important, since we developed microelectronics technology and computers, why should we procrastinate while other nations profit from our technology by applying it to the applications we should be undertaking?

REFERENCES

Gupton, James A., Jr., *Invention Disclosure*, Office of Patents and Copyright No. 087122, 1980.

Gupton, James A., Jr., *Microcomputers for External Device Control*, dilithium Press, Beaverton, Oreg., 1980, pp. 217–227.

Gupton, James A., Jr., "Nuclear Power Plant Emergency Damage Control Robot," *Robotic Age*, March/April 1983, pp. 18–21.

Mason, Carl P., *Medical Manipulator for Quadriplegic* (Abstract), Veterans' Administration Prosthetic Center, Rehabilitation Development Service, New York, 1979.

CHAPTER REVIEW QUESTIONS

1. What is a computer's post-X-ray diagnostic accuracy?

2. Define voice entry as it relates to computer data entry.

3. What is a voice-entry vocabulary?

4. Give four computer languages that are compatible with Shadow/Vet voice-entry commands.

5. What RAM memory is required for Shadow/Vet voice entry?

6. Why is a rechargeable gel battery specified for the source of mobile power for wheelchairs?

7. How are the wheel-driving motors coupled to a wheelchair's wheels?

8. Give the maximum speed-reduction ratio for a 10-in. 120-tooth driven gear when the the driver gear has a 0.176 in. OD and eight teeth.

9. Give at least four controllable wheelchair functions.

10. List the guidelines for any project development.

11. How must any loss function device appear to the person who has suffered the function loss?

12. Give four methods by which a quadriplegic can provide control inputs.

13. In the Veterans' Administration wheelchair manipulator prototype, what percentage of its functions were error-free?

14. How long will some radiation-contaminated parts of a nuclear power plant remain radioactive?

15. List three things about nuclear power plant construction that have been reported to be faulty.

16. Where should nuclear damage repair robots and repair materials be stored?

17. Why would a nuclear power plant repair robot have solid metal wheels?

18. Give five types of robot work hands that could be used in repairing ruptured metalwork.

19. What would be the best material to use as cutting teeth on the rotary cutter of the mine rescue robot?

20. What purpose do the perimeter water nozzles serve on the mine rescue robot?

21. How is the mine rescue robot steered?

22. What can cause ship hull ruptures in peacetime?

23. How does a wet metal deck affect a robot's wheel traction? A dual-track robot's function?

24. How may artificial gravity be produced on a spaceship in orbit?

25. On what is the ship damage repair robot's traction based?

26. What causes a ship to list?

27. What navigation assist would proximity sensors provide for a damage repair robot?

28. Would a fully computer controlled repair robot be practical for nuclear power plant or shipboard emergency damage control and repair?

29. List six immediate applications where robots could replace human workers in hazardous jobs.

30. Do we have the capability now to produce fully computerized and robotized manufacturing facilities?

GLOSSARY

Bayonet. A pointed or knifelike device attached to the barrel of a rifle and used to spear an enemy in hand-to-hand combat; a pointed attachment to any type of appendage that inserts into a mating receptacle.

Bulkhead. A naval term referring to the interior walls of a vessel.

Cyberphobia. Fear of computers.

Deactivate. To stop, halt, or turn off a device, process, or operation.

Deck. The horizontal covering of a ship's body or hull and the interior levels between the uppermost part of the hull and the ship's bottom; a naval term for floor.

Dogged. A term describing a method of latching or locking a hatch or door secured by wedging a metal lever into an angled metal receiver that forces the door closed to a watertight state.

Gel battery. A rechargeable battery using a semisolid or gel electrolyte, as opposed to the lead/acid-based electrolyte-type storage battery.

Hatch. A naval term referring to a deck or bulkhead opening which can be closed by a metal or wooden cover; means a door to the average person.

List. To lean or tilt from the normal vertical position. The list or lean of a ship resulting from hull damage flooding one side of the vessel.

Nozzle. A projecting spout attached to the end of a hose or pipe.

Overt. Obvious, manifest, or public.

Radioactive. A secondary source of lethal radiation formed by contact or exposure to nuclear radiation.

Sprocket gear. A form of traction gear with teeth designed to provide nonslip, positive traction between the sprocket gear and the driving chain link.

Winch. A motor-driven reel containing a length of flexible wire cable and a locking lever to prevent the reel from turning when power is turned off; a device for pulling heavy objects by means of a strong flexible cable.

Appendix A

National Semiconductor Corporation's CMOS Family*

Device	Function
MM54HC/74HC00	Quad 2-input NAND gate
MM54HC/74HC02	Quad 2-input NOR gate
MM54HC/74HC04	Hex inverter
MM54HC/74HCU04	Hex inverter (unbuffered)
MM54HC/74HC08	Quad 2-input AND gate
MM54HC/74HC10	Triple 3-input NAND gate
MM54HC/74HC11	Triple 3-input AND gate
MM54HC/74HC14	Hex inverting Schmitt trigger
MM54HC/74HC27	Triple 3-input NOR gate
MM54HC/74HC30	8-input NAND gate
MM54HC/74HC32	Quad 2-input OR gate
MM54HC/74HC42	BCD to decimal decoder
MM54HC/74HC51	Dual 2-input AND/OR/inverting gates
MM54HC/74HC58	Dual AND/OR gates
MM54HC/74HC73	Dual J-K flip-flop with clear
MM54HC/74HC74	Dual J-K flip-flop with preset and clear
MM54HC/74HC75	4-bit bistable latch with Q and Q output
MM54HC/74HC76	Dual J-K flip-flop with preset and clear
MM54HC/74HC85	4-bit magnitude comparator
MM54HC/74HC86	Quad 2-input exclusive-OR (XOR) gate
MM54HC/74HC107	Dual J-K flip-flop with clear
MM54HC/74HC109	Dual J-K flip-flop with preset and clear
MM54HC/74HC112	Dual J-K flip-flop with preset and clear

*Appendix A reprinted courtesy of National Semiconductor.

National Semiconductor Corporation's CMOS Family (cont.)

Device	Function
MM54HC/74HC113	Dual J-K flip-flop with preset
MM54HC/74HC123	Dual retriggerable monostable multivibrator
MM54HC/74HC125	Quad Tri-State buffer (low enable)
MM54HC/74HC132	Quad 2-input NAND Schmitt trigger
MM54HC/74HC133	13-input NAND gate
MM54HC/74HC137	1-to-8-line decoder with latch (inverting output)
MM54HC/74HC138	3-to-8-line decoder
MM54HC/74HC139	Dual 2-to-4-line decoder
MM54HC/74HC147	10-to-4-line priority encoder
MM54HC/74HC151	8-channel digital multiplexer
MM54HC/74HC153	Dual 4-input multiplexer
MM54HC/74HC154	4-to-16-line decoder
MM54HC/74HC157	Quad 2-input multiplexer
MM54HC/74HC158	Quad 2-input multiplexer (inverting output)
MM54HC/74HC160	Synchronous decade counter
MM54HC/74HC161	Synchronous binary counter
MM54HC/74HC162	Synchronous decade counter
MM54HC/74HC163	Synchronous binary counter
MM54HC/74HC164	8-bit serial in/parallel out shift register
MM54HC/74HC174	Hex flip-flop with clear
MM54HC/74HC175	Quad D-type flip-flop with clear
MM54HC/74HC181	4-bit arithmetic-logic unit
MM54HC/74HC182	Carry-look-ahead generator
MM54HC/74HC190	Up/down decade counter
MM54HC/74HC191	Up/down binary counter
MM54HC/74HC192	Synchronous decade up/down counter
MM54HC/74HC193	Synchronous binary up/down counter
MM54HC/74HC194	4-bit bidirectional universal shift register
MM54HC/74HC195	4-bit parallel shift register
MM54HC/74HC221	Dual monostable vibrator
MM54HC/74HC237	3-to-8-line decoder with address latches
MM54HC/74HC240	Inverting octal Tri-State buffer
MM54HC/74HC241	Octal Tri-State buffer (TTL buffer)
MM54HC/74HC242	Inverting quad Tri-State transceiver
MM54HC/74HC243	Quad Tri-State transceiver
MM54HC/74HC244	Octal Tri-State buffer
MM54HC/74HC245	Octal Tri-State transceiver
MM54HC/74HC251	8-channel Tri-State multiplexer
MM54HC/74HC253	Dual 4-channel Tri-State multiplexer
MM54HC/74HC257	Quad 2-channel Tri-State multiplexer
MM54HC/74HC259	8-bit addressable latch 3-to-8-line decoder
MM54HC/74HC266	Quad 2-input exclusive-NOR (XNOR) gate
MM54HC/74HC273	Octal D flip-flop
MM54HC/74HC280	9-bit off/even-parity generator/checker
MM54HC/74HC283	4-bit binary full adder
MM54HC/74HC292	Programmable 31/15-bit dividers/timers

National Semiconductor Corporation's CMOS Family (cont.)

Device	Function
MM54HC/74HC294	Programmable 31/15-bit dividers/timers
MM54HC/74HC298	Quad 2-channel with storage multiplexer
MM54HC/74HC299	8-bit Tri-State universal shift register
MM54HC/74HC353	8-channel Tri-State latched multiplexer
MM54HC/74HC356	8-channel Tri-State latched multiplexer
MM54HC/74HC365	Hex Tri-State buffer
MM54HC/74HC366	Inverting hex Tri-State buffer
MM54HC/74HC367	Hex Tri-State buffer
MM54HC/74HC368	Inverting Tri-State buffer
MM54HC/74HC373	Tri-State octal D-type latch
MM54HC/74HC374	Tri-State octal D-type flip-flop
MM54HC/74HC390	Dual 4-bit decade counter
MM54HC/74HC393	Dual 4-bit binary counter
MM54HC/74HC423	Dual nonretriggerable 1-shot multivibrator
MM54HC/74HC533	Tri-State octal D-type latch with inverting output
MM54HC/74HC534	Tri-State octal D-type flip-flop
MM54HC/74HC540	Octal Tri-State driver/buffer
MM54HC/74HC541	Octal Tri-State driver/buffer
MM54HC/74HC563	Tri-State octal D-type latch (inverting output)
MM54HC/74HC564	Tri-State octal D-type flip-flop with inverting output
MM54HC/74HC573	Tri-State D-type latch
MM54HC/74HC574	Tri-State octal D-type flip-flop
MM54HC/74HC595	8-bit parallel-to-serial shift register
MM54HC/74HC640	Inverting octal Tri-State transceiver
MM54HC/74HC643	Octal Tri-State transceiver
MM54HC/74HC646	Noninverting octal bus transceiver/register
MM54HC/74HC648	Inverting octal bus transceiver/register
MM54HC/74HC688	8-bit magnitude comparator (equality detector)
MM54HC/74HC942	300-baud modem
MM54HC/74HC4016	Quad bilateral analog switch
MM54HC/74HC4017	Decade counter/divider with 10 decoded outputs
MM54HC/74HC4020	14-stage binary counter
MM54HC/74HC4040	12-stage binary counter
MM54HC/74HC4046	Phase-locked loop
MM54HC/74HC4049	Hex inverting-logic-level down converter
MM54HC/74HC4050	Hex logic-level down converter
MM54HC/74HC4052	Dual 4-channel with storage multiplexer
MM54HC/74HC4053	Triple 2-channel with storage multiplexer
MM54HC/74HC4060	14-stage binary counter
MM54HC/74HC4066	Quad bilateral analog switch
MM54HC/74HC4075	Triple 3-input OR gate
MM54HC/74HC4078	8-input OR gate
MM54HC/74HC4316	Quad bilateral analog switch
MM54HC/74HC4351	8-channel analog multiplexer/demultiplexer with latch
MM54HC/74HC4514	4-to-16-line decoder with latch
MM54HC/74HC4518	Dual synchronous up counter

National Semiconductor Corporation's CMOS Family (cont.)

Device	Function
MM54HC/74HC4520	Dual synchronous up counter
MM54HC/74HC4538	Dual retriggerable monostable multivibrator
MM54HC/74HC4543	BCD-to-7-segment latch/decoder for liquid crystal display
MM54HC/74HC4560	NBCD adder

CMOS TTL Buffer Devices

MM54HCT/74HCT00	Quad 2-input NAND gate (TTL buffer)
MM54HCT/74HCT04	Hex inverter (TTL buffer)
MM54HCT/74HCT138	1-of-8 decoder (TTL buffer)
MM54HCT/74HCT240	Inverting octal Tri-State buffer (TTL buffer)
MM54HCT/74HCT241	Octal buffer Tri-State (TTL buffer)
MM54HCT/74HCT244	Inverting Tri-State buffer (TTL buffer)
MM54HCT/74HCT245	Octal transceiver Tri-State (TTL buffer)
MM54HCT/74HCT373	Octal D-type Tri-State latch (TTL buffer)
MM54HCT/74HCT374	Octal D-type Tri-State flip-flop (TTL buffer)
MM54HCT/74HCT640	Inverting octal Tri-State transceiver (TTL buffer)
MM54HCT/74HCT643	Octal Tri-State transceiver (TTL buffer)
MM54HCT/74HCT688	8-bit magnitude comparator (TTL buffer)

CMOS Interface and Microcomputer Devices

MM54C/74C00	Quad 2-input NAND gate
MM54C/74C02	Quad 2-input NOR gate
MM54C/74C04	Hex inverter
MM54C/74C08	Quad 2-input AND gate
MM54C/74C10	Triple 3-input AND gate
MM54C/74C14	Hex Schmitt trigger
MM54C/74C20	Dual 4-input NAND gate
MM54C/74C30	8-input AND gate
MM54C/74C32	Quad 2-input OR gate
MM54C/74C42	BCD-to-decimal decoder
MM54C/74C48	BCD-to-7-segment decoder driver
MM54C/74C73	Dual J-K flip-flop with clear
MM54C/74C74	Dual D flip-flop
MM54C/74C76	Dual J-K flip-flop with clear and preset
MM54C/74C83	4-bit binary full adder
MM54C/74C85	4-bit comparator
MM54C/74C86	Quad 2-input exclusive OR gate
MM54C/74C89	64-bit (16 × 4) RAM Tri-State
MM54C/74C90	4-bit decade counter
MM54C/74C93	4-bit binary counter
MM54C/74C95	4-bit right-shift/left-shift register
MM54C/74C107	Dual J-K flip-flop
MM54C/74C150	1-of-16 data selector
MM54C/74C151	4-bit data select/multiplexer with strobe
MM54C/74C154	4-to-16-line decoder

National Semiconductor Corporation's CMOS Family (cont.)

Device	Function
MM54C/74C157	Quad 2-input multiplexer
MM54C/74C160	Decade counter with asynchronous clear
MM54C/74C161	Binary counter with asynchronous clear
MM54C/74C162	Synchronous decade counter
MM54C/74C163	Synchronous binary counter
MM54C/74C164	8-bit serial-in/parallel-out shift register
MM54C/74C165	8-bit parallel-in/serial-out shift register
MM54C/74C173	Tri-State quad D flip-flop
MM54C/74C174	Hex D flip-flop
MM54C/74C175	Quad D flip-flop
MM54C/74C192	Decade up/down counter
MM54C/74C193	Binary up/down counter
MM54C/74C195	4-bit parallel access shift register
MM54C/74C200	256-bit RAM Tri-State
MM54C/74C221	Dual monostable multivibrator
MM54C/74C240	Tri-State octal buffer (inverting output)
MM54C/74C244	Tri-State octal buffer
MM54C/74C373	Octal flow-through latch Tri-State
MM54C/74C374	Octal D-type flip-flop Tri-State

CMOS TTL Interface

Device	Function
MM54C/74C901	Hex inverting buffer (TTL interface)
MM54C/74C902	Hex noninverting buffer (TTL interface)
MM54C/74C903	Hex inverting buffer (TTL interface)
MM54C/74C904	Hex noninverting buffer (TTL interface)
MM54C/74C905	12-bit successive approximation register
MM54C/74C906	Open drain buffer (active pull down)
MM54C/74C907	Open drain buffer (active pull down)
MM54C/74C908	Dual high-voltage driver
MM54C/74C909	Quad comparator
MM54C/74C910	256-bit (64×4) RAM Tri-State
MM54C/74C911	4-digit LED display controller
MM54C/74C912	6-digit LED display controller
MM54C/74C914	Hex Schmitt trigger with extended input voltage
MM54C/74C915	7-segment top BCD encoder
MM54C/74C922	16-key keyboard encoder
MM54C/74C923	16-key keyboard encoder
MM54C/74C929	1024-bit static RAM
MM54C/74C930	1024-bit static RAM
MM54C/74C932	Phase comparator
MM54C/74C933	Address bit comparator
MM54C/74C941	Tri-State octal buffer
MM54C/74C945	4-digit up/down counter/latch/decoder driver
MM54C/74C946	$4\frac{1}{2}$-digit counter decoder driver
MM54C/74C947	4-digit up/down counter/latch/decoder driver

National Semiconductor Corporation's CMOS Family (cont.)

Device	Function
MM54C/74C949	8-bit microprocessor-compatible A/D converter
MM54C/74C989	64-bit (16 × 4) RAM Tri-State, +5 V

Oscillator Dividers, Real-Time Clocks, and Display Drivers

Device	Function	Performance
MM5368	Oscillator/divider	Input frequency 32.768 Hz; output frequency 1 Hz, 10 Hz, 50/60 Hz
MM5369	Oscillator/divider	Input frequency 3.58 MHz; output frequency 50 Hz, 60 Hz, 100 Hz selectable
MM53107	Oscillator/divider	Input frequency 2.097152 MHz; output frequency 100 Hz
MM5452/53	32/33-bit LCD display driver	Direct drive/cascade capability/ alphanumeric
MM5483	31-bit LCD display driver	Direct drive/cascade capability/ alphanumeric
MM58167A	CPU real-time clock	8-bit data bus/10th of sec/min/day of week/month/year
MM58174A	CPU real-time clock	4-bit data bus/10th of sec/min/day of week/month/year
MM58201	192-bit LCD display driver	Drives up to 8 backplanes and 24 columns
MM58241/8	32/35-bit VF display driver	Direct interface to 60-V VF display
MM58341/8	32/35 bit VF display driver	Direct interface to 32-V VF display
MM58538	32-bit LCD display driver	Direct drive/cascade operation/ alphanumeric
MM58538/9	MUX LCD display driver	External refresh/drives up to 8 back-planes and 26 columns
MM58540/8	MUX LCD display driver	Drives 32 columns or 32 rows, cascade capability
MM58274	MPU real-time clock	10th of sec to 10's of years; 12/24-hour output, low power standby operation
MM78/88C29	Quad single-end line driver	Wide V_{cc} (3 to 15 V) lower resistance, 20 Ω
MM78/88C30	Dual differential line driver	Wide V_{cc} (3 to 15 V) lower resistance, 20 Ω

CMOS Microprocessors and Peripherals

Device	Function	Clock frequency (MHz)	OP (mA)
NSC800-1	8-bit CPU	1	10
NSC800D/883B	8-bit CPU	2.5	15
NSC800-4	8-bit CPU	4	21
NSC810-1	128-byte static RAM counters/timers	1	6
NSC810D/883B	128-byte static RAM counters/timers	2.5	10
NSC810-4	128-byte static RAM counters/timers	4	12
NSC830-1	2-kilo byte ROM, I/O	1	6
NSC830D/883B	2-kilo byte ROM, I/O	2.5	10
NSC830-4	2-kilo byte ROM, I/O	4	12
NSC831-1	I/O	1	6
NSC831D/883B	I/O	2.5	10
NSC831-4	I/O	4	12

Appendix B

Decimal-to-Metric Conversions

Decimal	Millimeters $(10^{-3}$ m)	Microns $(10^{-6}$ m)	Nanometers $(10^{-9}$ m)	Angstroms $(10^{-10}$ m)
1.0	25.4	25,400		254,000
0.001	0.0254	25.4		
0.0009	0.02286	22.86		
0.0008	0.02032	20.32		
0.00079	0.02000	20.00		
0.00075	0.01905	19.05		
0.00075−	0.01900	19.00		
0.00071	0.01800	18.00		
0.0007	0.01778	17.78		
0.00067	0.01700	17.00		
0.00065	0.01651	16.51		
0.00063	0.01600	16.00		
0.0006	0.01524	15.25		
0.00059	0.01500	15.00		
0.00055	0.01400	14.00		
0.00051	0.01300	13.00		
0.0005	0.01270	12.70		
0.00047	0.01200	12.00		
0.00043	0.01100	11.00		
0.0004	0.01016	10.16		
0.00039	0.01000	10.00	10,000	100,000

Decimal	Millimeters $(10^{-3}\ m)$	Microns $(10^{-6}\ m)$	Nanometers $(10^{-9}\ m)$	Angstroms $(10^{-10}\ m)$
0.00035	0.00900	9.00		
0.0032	0.00800	8.00		
0.0003	0.00762	7.62		
0.00028	0.00700	7.00		
0.00025	0.00635	6.35		
0.00024	0.00600	6.00		
0.0002	0.00508	5.08		
0.00019	0.00500	5.00	5,000	50,000
0.00016	0.00400	4.00	4,000	40,000
0.00015	0.00381	3.81	3,810	38,100
0.00012	0.00300	3.00	3,000	30,000
0.0001	0.00254	2.54	2,540	25,400
0.000098		2.50	2,500	25,000
0.000091		2.41	2,410	24,100
0.000090		2.30	2,300	23,000
0.000087		2.286	2,286	22,860
0.000083		2.100	2,100	21,000
0.000080		2.032	2.032	20,320
0.000079		2.000	2,000	20,000
0.000075		1.900	1,900	19,000
0.000071		1.800	1,800	18,000
0.000070		1.778	1,778	17,780
0.000067		1.700	1,700	17,000
0.000063		1.600	1,600	16,000
0.000060		1.524	1,524	15,240
0.000059		1.500	1,500	15,000
0.000055		1.400	1,400	14,000
0.000051		1.300	1,300	13,000
0.000050		1.270	1,270	12,700
0.000047		1.200	1,200	12,000
0.000043		1.100	1,100	11,000
0.000040		1.016	1,060	10,600
0.000030		1.000	1,000	10,000
0.000035		0.900	900	9,000
0.000032		0.800	800	8,000
0.000030		0.762	760	7,600
0.000028		0.700	700	7,000
0.000024		0.600	600	6,000
0.000020		0.508	508	5,800
0.000019		0.500	500	5,000
0.000016		0.400	400	4,000
0.000015		0.381	381	3,810
0.000012		0.300	300	3,000
0.000010		0.254	250	2,500
0.000009		0.2286	228.6	2,286
0.000001		0.0254	25.4	254

Powers of 10	Prefix	Symbol
10^{12}	tera	T
10^{9}	giga	G
10^{6}	mega	M
10^{3}	kilo	k
10^{2}	hekto	h
10	deka	da
10^{-1}	deci	d
10^{-2}	centi	c
10^{-3}	milli	m
10^{-6}	micro	μ
10^{-9}	nano	n
10^{-12}	pico	p
10^{-15}	femto	f
10^{-18}	alto	a

Appendix C

CCD4001
Robotics Camera
Automation Camera Series

CCD Imaging

DESCRIPTION

The CCD4001 Robotics Camera is a small, rugged, solid-state camera designed for use in industrial environments. The CCD4001 incorporates a 256 × 256 element sensor with a square pixel pitch format. The camera can provide NTSC televison video output signals for display of images on standard monitors or for digital analysis using NTSC image processing equipment. The camera output is a 525-line, interlaced format with a resolution of 256 lines per field by 256 elements per line. Each frame of video is composed of two identical fields.

The camera can be used as a single piece unit or separated into a camera control unit and sense head connected by a flexible cable. The small, rugged, lightweight sense head is designed to tolerate high accelerations and vibrations which, for example, might be encountered on a quickly moving robot arm.

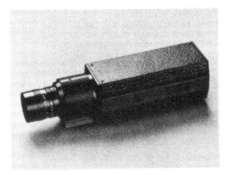

FEATURES

- SMALL, COMPACT ENCLOSURES WITH SEALED REMOTABLE SENSE HEAD
- WELL SUITED FOR USE IN RUGGED INDUSTRIAL ENVIRONMENTS
- 256 × 256 ELEMENT NON-INTERLACED SENSOR
- SQUARE PIXEL PITCH FORMAT
- ELEMENT BLOOMING CONTROL
- ALL SOLID-STATE, BURIED CHANNEL CCD
- SELF-CONTAINED
- NO LAG OR GEOMETRIC DISTORTION
- WIDE DYNAMIC RANGE: TYPICALLY 1000:1
- ELECTRONICALLY VARIABLE FRAME RATES
- GEN-LOCK CAPABILITY
- OPTIONAL EXTERNAL TIMING CONTROL

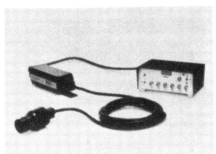

CCD4001 BLOCK DIAGRAM

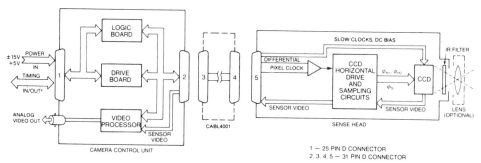

1 — 25 PIN D CONNECTOR
2, 3, 4, 5 — 31 PIN D CONNECTOR

* OPTIONAL USAGE

CCD4001

SPECIFICATIONS

Sensor Scanning Format — Non-interlaced 256 elements per line. 256 lines per frame.

Sensor Scan Timing — Field rate is 60Hz (using internal clock).

Camera Output Format — Interlaced 2 identical fields per frame.

Camera Output Timing — NTSC compatible frame rate is 30Hz, data rate is 6.14M elements per second under control of an internal, crystal-controlled oscillator. (External clock input may be accepted for variable frame rates.)

Synchronization — May be gen-locked with horizontal and vertical drive signal inputs. Also the scan rate may be controlled by clock input; frames can be synchronized by field index input.

Output Signals — Analog Video: 1.0V p-p Non-composite, 75 ohm, Black = 0 ± 0.1V (Sync may be added with pc strap.) Timing: Vertical and Horizontal Drive, Composite Sync and Blanking, Frame Index, Data Rate Clock, Video Valid.

Resolution — 256 lines per picture height, 256 lines per picture width.

Sensor — Monolithic silicon CCD. Element spacing: 22μm C-C horizontal; 22μm C-C vertical. Aspect Ratio: 1:1 (horizontal: vertical). Image Diagonal: 7.94mm.

Input Signals — (At user's option) Timing: Vertical and Horizontal Drive, Timing Mode Selects, Master Clock, Field Index.

Dynamic Range — 1000:1.

Saturation Irradiance — 6.0μw/cm^2 at normal scan rate.

Lens — Camera accepts C-mount lenses, 1" vidicon types are recommended. (See Options)

Cosmetic Performance* (At 25°C)

Dark Signal Shading Non-Uniformity (DSSNU) ≤20 mVp-p (≤2% Peak Output).

Photoresponse Shading Non-Uniformity (PRSNU) ≤5% of V$_{OUT}$.

Largest Dimension of Blemished Area ≤3 Contiguous Elements

Number of Blemished Elements ≤100 pixels.

Number of Blemished Columns = 0

Enclosure — Sense head is gasket and O-ring sealed.

Dimensions — See Figure.

Weight — Sense Head: ≤8 oz., Sense Head and Control Unit: 2.2 lbs.

Environmental Conditions — Operating Ambient Temperature: Sense Head: 0-50°C. Control Unit: 0-50°C. Acceleration and Shock (Sense Head): >100G, any axis. Vibration: 0-2000Hz, 2G, any axis.

Power Requirements — 5W input, ±15, +5Vdc.

*Notes

1. These characteristics are measured at uniform illumination levels providing video signal output levels from 0V (Black) to 1.0Vp-p (Full White).

2. Blemished elements are contained in randomly located small (not larger than 3 × 3 elements) areas where the video output signal differs from the output of the surrounding area by >±100 mV.

3. Certain video anomalies may sometimes be observed in the displayed camera output when the sensor illumination level exceeds the light level needed to achieve the normal camera saturation output video signal level of 1.0Vp-p.

4. The amplitude of DSSNU and the output of all sensor elements in the dark should be expected to double for each 5-10 degree C increase in sense head temperature.

5. Cameras with improved or degraded cosmetic performance specifications are available to volume purchasers with price adjustments. Please consult the factory for more information.

CCD4001

FUNCTIONAL DESCRIPTION

The CCD4001 camera is shipped as a single-piece unit comprised of a sense head and camera control unit connected by a dovetail member. These two subunits may be separated by a 12-foot cable (included with the camera) permitting remote operation of the sense head. A Power-Supply Unit is also included with the CCD4001 camera.

Each camera subunit contains electronics as illustrated in the block diagram.

Image Sensor

The image detector used in the CCD4001 camera sense head (See Block Diagram) is a selected, monolithic 256 × 256-element charge-coupled device (CCD) image sensor. The buried channel CCD architecture employed in the sensor minimizes noise and allows high frame rates without sacrificing charge transfer efficiency. The advanced Fairchild CCD technology allows the camera to offer low lag and geometric distortion, lower power consumption, small size, and unusual robustness for use in industrial environments.

Sense Head

The sense head contains the image sensor and circuitry for generating the high frequency horizontal register and sample pulse clock signals for CCD control and a buffer for the sensor video. All other CCD timing and drive electronics, supply and bias voltages as well as video processing electronics are contained in the camera control unit. Light reaching the sensor is filtered by a 2.0 mm thick Schott BG-38 glass in order to eliminate IR content and give a near photopic spectral sensitivity. The sensor is rigidly held in position and precisely aligned with respect to the sense head mounting foot. The sensor is thermally connected to the exterior walls at the sense head by low thermal resistance hard-anodized aluminum internal structures. The sense head is sealed by O-rings, gaskets and by bonding of the filter glass into the lens mount.

Camera Control Unit

The camera control unit houses three pc cards performing the functions of camera and sensor timing control, CCD drive and video processing, interconnected by two pc mother boards. Camera timing is controlled by an internal 12.285MHz master clock oscillator located on the drive board. The frequency of this oscillator is controlled by a phase locked loop circuit when the camera timing is "gen-locked" to external vertical and horizontal drive signals. The data rate in pixels per second is the master clock frequency divided by 2.

Timing waveforms for sensor drive and sync signals for the formation of NTSC compatible video are derived from the master clock signal. These timing signals are then fed to the drive board where the TTL level signals are altered to CCD drive level signals required to operate the sensor. From the drive board, sensor clocks are fed through the 31-pin D connector to the sense head and the composite sync signal is forwarded to the video processor board.

The video processor receives sensor video from the CCD in the sense head. The video is line clamped, amplified and blanked with composite blanking from the logic board, and may be summed with a composite sync signal from the drive board to yield RS170 composite video at the BNC output at the back of the camera. An Automatic Video Gain Control circuit may be actuated by connection of pin 18 to pin 19; about 20 db of gain increase is available.

Power Supply Unit

The cameras require inputs of ±15 and +5Vdc which are provided by the Power Supply Unit. Regulators on-board provide all voltage levels needed to drive the amplifiers and various clocks. The power supply unit voltages are derived from power line voltages of 120±10 or 240±20Vac, 47−63Hz. The front panel of the power supply unit provides BNC-connector access to composite blanking, composite sync, vertical drive and horizontal drive input and output signals and an external clock input as TTL levels. The rear panel contains connectors for interfacing the camera and power supply to the VIP100 Vision Interface Processor. A 6′ (approx. 2m) cable is provided for interconnection of the camera control unit and the power supply.

OPTIONS AND ORDER INFORMATION

Order Code	Description
CCD4001	Includes camera, power supply unit, interconnect cable and remote sense head cable.
CAM4001	CCD4001,less power supply unit, interconnect cable and sense head cable.
PWRSPLY	Power Supply Unit for CAM4001
CABL4001	Remote Sense Head Cable for CAM4001
Lenses	1″ Vidicon type C-mount lenses are available in focal lengths of 13mm, 25mm and 50mm.
Monitor	NTSC monitor

TYPICAL PERFORMANCE CURVES

NORMALIZED SPECTRAL RESPONSE

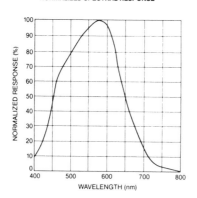

HORIZONTAL CONTRAST TRANSFER FUNCTION

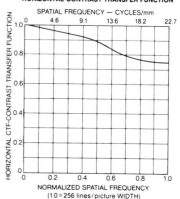

VERTICAL CONTRAST TRANSFER FUNCTION

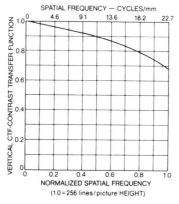

I/O PIN CONNECTIONS

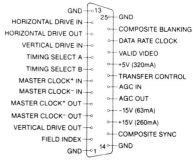

MATING CONNECTOR IS TYPE DB-25S
BY TRW OR EQUIVALENT

CCD4001

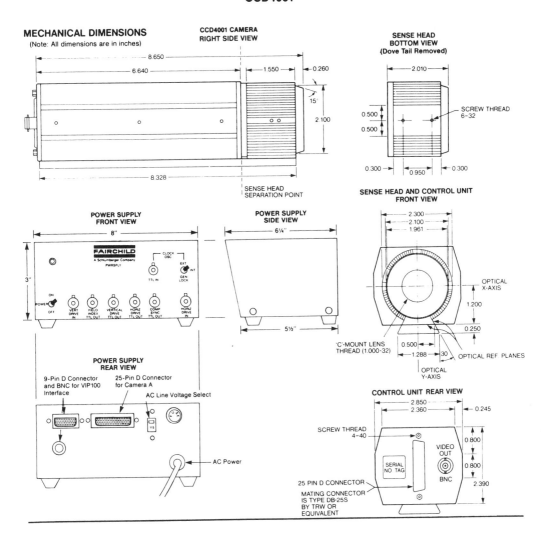

MECHANICAL DIMENSIONS
(Note: All dimensions are in inches)

CCD4001 CAMERA
RIGHT SIDE VIEW

8.650
6.640
1.550
0.260
15°
2.100
8.328

SENSE HEAD
SEPARATION POINT

SENSE HEAD
BOTTOM VIEW
(Dove Tail Removed)

2.010
0.500
0.500
SCREW THREAD
6–32
0.300   0.950   0.300

POWER SUPPLY
FRONT VIEW

8"
3"

FAIRCHILD
A Schlumberger Company
PWRSPLY

CLOCK OSC
TTL IN
EXT
GEN LOCK
INT

POWER
ON
OFF

VERT DRIVE IN
FIELD INDEX TTL OUT
VERTICAL DRIVE TTL OUT
HORIZ DRIVE TTL OUT
COMP SYNC TTL OUT
HORIZ DRIVE IN

POWER SUPPLY
SIDE VIEW

6¼"
5½"

SENSE HEAD AND CONTROL UNIT
FRONT VIEW

2.300
2.100
1.961

OPTICAL
X-AXIS
1.200
0.250

'C'-MOUNT LENS
THREAD (1.000-32)

0.500
1.288
30
OPTICAL REF PLANES

OPTICAL
Y-AXIS

POWER SUPPLY
REAR VIEW

9-Pin D Connector
and BNC for VIP100
Interface

25-Pin D Connector
for Camera A

AC Line Voltage Select

115

AC Power

CONTROL UNIT REAR VIEW

2.850
2.360
0.245

SCREW THREAD
4-40

0.800
0.800
2.390

VIDEO
OUT

SERIAL
NO TAG

BNC

25 PIN D CONNECTOR

MATING CONNECTOR
IS TYPE DB-25S
BY TRW OR
EQUIVALENT

Appendix D

The stepper motor is a device used to convert electrical pulses into discrete mechanical rotational movements.

The Airpax Corporation stepper motors described in this handbook are two-phase permanent magnet motors which provide discrete angular movement every time the polarity of a winding is changed.

CONSTRUCTION

In a typical motor, electrical power is applied to two coils. Two stator cups formed around each of these coils, with pole pairs mechanically displaced by ½ a pole pitch, become alternately energized North and South magnetic poles. Between the two stator-coil pairs the displacement is ¼ of a pole pitch.

The permanent magnet rotor is magnetized with the same number of pole pairs as contained by one stator-coil section.

Interaction between the rotor and stator (opposite poles attracting and likes repelling) causes the rotor to move ¼ of a pole pitch per winding polarity change. A two-phase motor with 12 pole pairs per stator-coil section would thus move 48 steps per revolution or 7.5° per step.

ELECTRICAL INPUT

The normal electrical input is a four-step switching sequence as is shown in Figure 2.

Continuing the sequence causes the rotor to rotate forward. Reversing the sequence reverses the direction of rotation. Thus, the stepper motor can be easily controlled by a pulse input drive which can be a two flip-flop logic circuit operated either open or closed loop. Operated at a

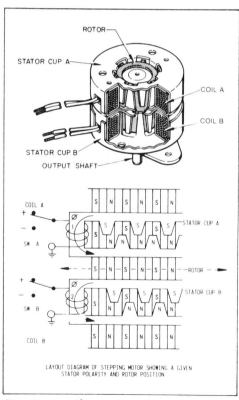

LAYOUT DIAGRAM OF STEPPING MOTOR SHOWING A GIVEN STATOR POLARITY AND ROTOR POSITION

Fig. 1 Cutaway 2 Ø Permanent Magnet Stepper Motor.

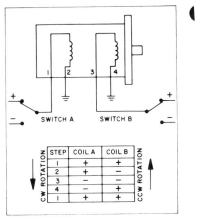

Fig. 2 Schematic — 4-Step Switching Sequence.

fixed frequency, the electrical input to the motor is a two-phase 90° shifted square wave as shown below.

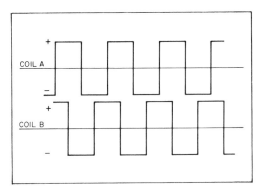

Fig. 3 Voltage Wave Form — Fixed Frequency — 4-Step Sequence.

Since each step of the rotor can be controlled by a pulse input to a drive circuit, the stepper motor used with modern digital circuits, micro-processors and transistors provides accurate speed and position control along with long life and reliability.

STEP ANGLE

Step angles for steppers are available in a range from .72° to 90°. Standard step angles for Airpax Corporation Steppers are:

    7.5° — 48 steps per rev.
    15° — 24 steps per rev.
    18° — 20 steps per rev.

A movement of any multiple of these angles is possible. For example, six steps of a 15° stepper motor would give a movement of 90°.

ACCURACY

The no load or constant load accuracy of each step is within ±6.5%, noncumulative. Therefore, a 7.5° stepper motor will position to within 0.5°, whether the rotational movement is 7.5° — one step, or 7,500° — one thousand steps.

The step error is noncumulative. It averages out to zero within a 4-step sequence which corresponds to 360 electrical degrees. A particular step condition of the 4-step sequence repeatedly uses the same coil, magnetic polarity and flux path. Thus, the most accurate movement would be to step in multiples of four since electrical and magnetic inbalances are eliminated. Increased accuracy also results from movements which are multiples of two steps. Keeping this in mind, positioning applications should use 2 or 4 steps (or multiples thereof) for each desired measured increment, wherever possible.

TORQUE

The torque produced by a specific stepper motor depends on several factors:

    1/ The Step Rate
    2/ The Drive Current Supplied to the Windings
    3/ The Drive Design

HOLDING TORQUE

At standstill (zero steps/sec. and rated current) the torque required to deflect the rotor a full step is called the Holding Torque. Normally, the holding torque is higher than the running torque and thus acts as a strong brake in holding a load. Since deflection varies with load, the higher the holding torque the more accurate the position will be held. Note in the curve below that a two step deflection corresponding to a phase displacement of 180° results in zero torque. A one step plus or minus displacement represents the initial lag that occurs when the motor is given a step command.

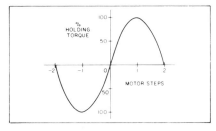

Fig. 4 Torque Deflection.

RESIDUAL TORQUE

The non-energized detent torque of a PM stepper motor is called residual torque. A result of the permanent magnet flux and bearing friction, it has a value of approximately 1/10 the holding torque. This characteristic of PM steppers is useful in holding a load in the proper position even when the motor is de-energized. The position, however, will not be held as accurately as when the motor is energized.

DYNAMIC TORQUE

A typical torque versus step rate characteristic curve is shown in Figure 5.

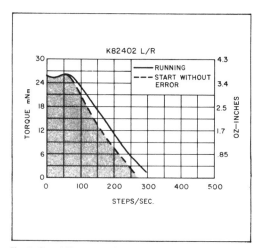

Fig. 5   Speed/Torque — (Airpax K82402 L/R Stepper).

The **Start Without Error** curve shows what torque load the motor can start and stop without loss of a step when started and stopped at a constant step or pulse rate.

The **Running** curve is the torque available when the motor is slowly accelerated to the operating rate. It is thus the actual dynamic torque produced by the motor. This curve is sometimes called the slew curve.

The difference between the Running and the Start Without Error torque curves is the torque lost due to accelerating the motor rotor inertia.

The speed-torque characteristic curves are the key to selecting the right motor and the control drive method for a specific application.

In order to properly analyze application requirements, the load torque must be defined as being either Frictional and/or Inertial. A "Handy Formula" section in this handbook on pages 12 and 13 may assist you in resolving the load torque values. Also, an additional "Application Notes" section is located on pages 10 and 11.

Use the Start Without Error curve if the control circuit provides no acceleration and the load is frictional only.

Applications where:

No acceleration — Frictional Load

Example: Frictional Torque Load.

Using a torque wrench, a frictional load is measured to be 10.6 mNm (1.5 oz-in). It is desired to move this load 67.5° in .06 sec. or less.

Solution:

1. If a 7.5° motor is used, then the motor would have to take 9 steps to move 67.5°.

    A rate of $v = \frac{9}{.06} = 150$ step/sec. or higher is thus required.

2. Referring to Fig. 6, the maximum Start Without Error rate with a torque of 10.6 mNm is 170 steps/sec. (It is assumed no acceleration control is provided).

3. Therefore, a K82402 motor could be used at 150 steps per second — allowing a safety factor.

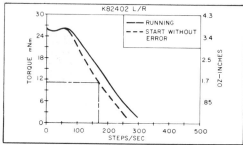

Fig. 6 Speed/Torque — Frictional Load.

Use the Running curve, in conjunction with a Torque = Inertia x Acceleration equation ($T = J\alpha$), when the load is inertial and/or acceleration control is provided.

In this equation, acceleration or ramping $\alpha = \frac{\Delta v}{\Delta t}$ is in radians/sec².

RAMPING

Acceleration control or ramping is normally accomplished by gating on a voltage controlled oscillator and associated charging capacitor. Varying the RC time constant will give different ramping times. A typical VCO acceleration control frequency plot for an incremental movement with equal acceleration and deceleration time would be as shown in Fig. 7.

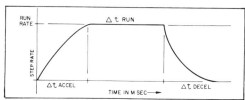

Fig. 7 Step Rate/Time.

Acceleration may also be accomplished by dividing the frequency. For example, the frequency could start at a ¼ rate, go to a ½ rate, ¾ rate and finally the running rate.

A. Applications where:

Ramping acceleration or deceleration control time allowed:

$$T_J (\text{Torque mNm}) = J_T \times \frac{\Delta v}{\Delta t} \times K$$

Where J_T = Rotor Inertia (g.m²) plus Load Inertia (g.m²)

Δv = Step rate change
Δt = Time allowed for acceleration in seconds
$K = \dfrac{2\pi}{\text{steps/rev}}$ (converts steps/sec to radians/sec)
K = .13 for 7.5° — 48 steps/revolution
K = .26 for 15° — 24 steps/revolution
K = .314 for 18° — 20 steps/revolution

In order to solve an application problem using acceleration ramping, it is usually necessary to make several estimates according to a procedure similar to the one used to solve the following example.

Example: Frictional Torque plus Inertial Load with Acceleration Control.

An assembly device must move 4 mm in less than 0.5 sec. The motor will drive a lead screw through a gear ratio. The lead screw and gear ratio were selected so that 100 steps of a 7.5° motor = 4 mm. The total Inertial Load (rotor + gear + screw) = 25 x 10⁻⁴ g.m². The Frictional Load = 6 mNm

Solution:

1. Select a stepper motor running curve which allows a torque in excess of 6 mNm at a step rate greater than

$$v = \frac{100 \text{ steps}}{0.5 \text{ sec}} = 200 \text{ steps/sec.}$$

Referring to Fig. 8, determine the maximum possible rate (vF) with the frictional load only.

2. Make a first estimate of a working rate (a running rate less than the maximum) and determine the torque available to accelerate the inertia (excess over T_F).

$$T_1 - T_F = 10 - 6 = 4 \text{ mNm}$$

(torque available for acceleration at 210 steps/sec).

3. Using a 60% safety factor

$$(4 \text{ mNm} \times .6 = 2.4 \text{ mNm}),$$

calculate Δt to accelerate. (Refer to Fig. 7).

From the $T_J = J_T \times \dfrac{\Delta v}{\Delta t} \times K$ equation,

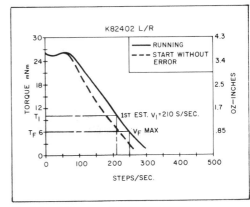

Fig. 8 Speed/Torque — Friction Plus Inertia.

$$2.4 \text{ mNm} = \frac{25 \times 10^{-4} \times 210 \times .13}{\Delta t}$$

Therefore to accelerate Δt = .028 sec

Note: The same amount of time is allowed to decelerate.

4. The number of steps used to accelerate and decelerate,

$$N_A + N_D = \frac{v}{2} \Delta t \times 2$$

or $N_A + N_D = v\Delta t$

$$= 210 (.03) = 6 \text{ steps}$$

5. The time to move at the run rate

$$\Delta t_{run} = N_T - (N_A + N_D) = \frac{100-6}{210} = .447 \text{ sec}$$

Where N_T = Total move of 100 steps

6. The total time to move is thus

$$\Delta t_{run} + \Delta t_{accel} + \Delta t_{decel}$$

$$.447 + .028 + .028 = 0.5 \text{ sec}$$

This is the first estimate. You may make the move slower if more safety is desired, or faster if you want to optimize it. At this time, you may wish to consider a faster motor drive combination as will be discussed on page 8.

B. Applications where:

No ramping acceleration or deceleration control time allowed.

Even though no acceleration time is provided, the stepper motor can lag a maximum of 2 steps or 180 electrical degrees. If the motor goes from zero steps/sec to v steps/sec, the lag time Δt would be $\dfrac{2}{v}$ sec.

Thus the torque equation for no acceleration or deceleration is:

$$T \text{ (Torque mNm)} = J_T \times \frac{v^2}{2} \times K$$

Where:

J   Rotor Inertia (g.m²) plus Load Inertia (g.m²)

$$v = \text{Steps/sec rate}$$
$$K = \frac{2\pi}{\text{step/rev}}$$

("K" values as shown in application A on page 4)

Example: Friction plus Inertia — No acceleration ramping.

A tape capstan is to be driven by a stepper motor. The frictional drag torque (T_F) is 8.6 mNm and the Inertia of the capstan is 8×10^{-4} g.m². The capstan must rotate in 7.5° increments at a rate of 170 steps per second.

Solution:

Since a torque greater than 8.6 mNm at 170 steps per second is needed, consider a K82402 motor.
The Total Inertia = Motor Rotor Inertia + Load Inertia.

$$J_T = J_R + J_L$$
$$= (10 \times 10^{-4} + 8 \times 10^{-4}) \text{ g.m}^2$$
$$= 18 \times 10^{-4} \text{ g.m}^2$$

1. Since there is no acceleration ramping, use the equation:

$$T_J = J_T \times \frac{v^2}{2} \times K \qquad (K = .13)$$

$$T_J = 18 \times 10^{-4} \times \frac{170^2}{2} \times .13$$

$$T_J = 3.4 \text{ mNm}$$

2. Total Torque $= T_F + T_J$

$$= 8.6 + 3.4$$

$$= 12 \text{ mNm}$$

3. Refer to the running curve Fig. 9, at 170 steps per second, the available torque is 15 mNm. Therefore, the K82402 motor can be used with a safety factor.

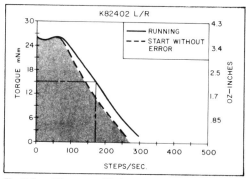

Fig. 9 Speed/Torque — Friction Plus Inertia.

STEP FUNCTION — SINGLE STEP

When a single step of a motor is made, a typical response is as shown in Figure 10.

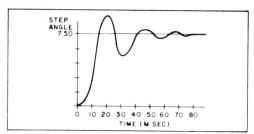

Fig. 10 Single Step Response.

The actual response for a given motor is a function of the power input provided by the drive and the load. Increasing the frictional load or adding external damping can thus modify this response if it is required.

Mechanical dampers such as slip pads or plates, or devices such as a fluid coupled flywheel can be used but add to system cost and complexity. Electronic damping can also be accomplished. A first time delay and a reverse pulse and then a second time delay and forward pulse is added to every move pulse or to the last pulse in a movement. Delaying the final pulse of an incremental movement can also be used to effect damping.

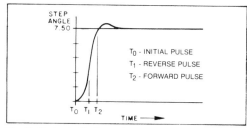

Fig. 11 Electronically Damped Response.

STEP FUNCTION — MULTIPLE STEPPING

Multiple stepping can offer several alternatives. A 7.5° motor moving 12 steps or a 15° motor moving 6 steps to give a 90° output move would have less overshoot, be stiffer, and relatively more accurate than a motor with a 90° step angle. Also, the pulses can be timed to shape the velocity of the motion; slow during start, accelerate to maximum velocity, then decelerate to stop with minimum ringing.

RESONANCE

If a stepper motor is operated no-load over the entire frequency range, one or more natural oscillating resonance points may be detected either audibly or by vibration sensors. Some applications may be such that operation at these frequencies should be avoided or external damping, added inertia, or a softer drive used. A permanent magnet stepper motor, however, will not exhibit the instability and loss of steps often found in variable reluctance stepper motors since the PM has a higher rotor inertia and a stronger detent torque.

DRIVE METHODS

The normal drive method, as previously stated, is the 4-step sequence shown in Fig. 2; however, the following methods are also possible.

WAVE DRIVE

Energizing only one winding at a time, as is indicated in Fig. 12 is called Wave Excitation. It produces the same increment as the four-step sequence.

Since only one winding is on, the hold and running torque with rated voltage applied will be reduced 30%. Within limits, the voltage can be increased to bring output power back to near rated torque value. The advantage of this type of drive is increased efficiency while the disadvantage is decreased step accuracy.

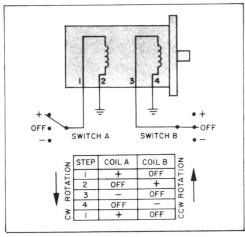

Fig. 12 Schematic — Wave Drive Switching Sequence.

HALF STEP

It is also possible to step the motor according to an eight step sequence to obtain a half step — such as a 3.75° step from a 7.5° motor.

Applications utilizing this should be aware of the fact that the holding torque will vary for every other step since only one winding will be energized for a step position but on the next step two windings are energized. This gives the effect of a strong step and a weak step. Also, since the winding and flux conditions are not similar for each step when ½ stepping, accuracy will not be as good as when full stepping.

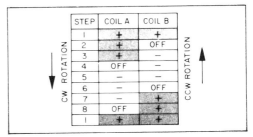

Fig. 13 Half Step or 8 Step Switching Sequence.

BIPOLAR AND UNIPOLAR OPERATION

All Airpax stepper motors are available with either 2 coil bipolar, or 4 coil unipolar windings.

The stator flux with a BIPOLAR winding is reversed by reversing the current in the winding. It requires a push-pull bipolar drive as shown in Fig. 14. Care must be taken to design the circuit so that the transistors in series do not short the power supply by coming on at the same time. Properly operated, the bipolar winding gives the optimum motor performance at low to medium step rates.

A UNIPOLAR winding has 2 coils wound on the same bobbin per stator half. Flux is reversed by energizing one coil or the other coil from a single power supply. The use of a unipolar winding, sometimes called a bifilar winding, allows the drive circuit to be simplified. Not only are half as many power switches required (4 vs. 8), but the timing is not as critical to prevent a current short through two transistors as is possible with a bipolar drive.

For a unipolar motor to have the same number of turns per winding as a bipolar motor, the wire diameter must be decreased and therefore the resistance increased. As a result unipolar motors have 30% less torque at low step rates. However, at higher rates the torque outputs are equivalent.

HIGHER PERFORMANCE

A motor operated at a fixed rated voltage has a decreasing torque curve as the frequency or step rate increases. This is due to the fact that the rise time of the coil limits the percentage of power actually delivered to the motor. This effect is governed by the inductance to resistance ratio of the circuit (L/R).

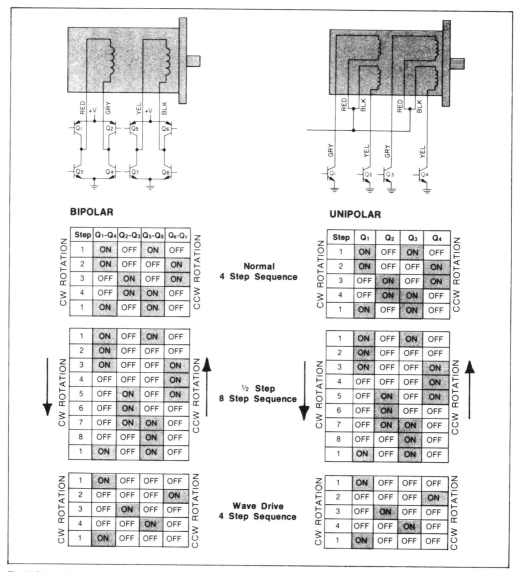

Fig. 14 Schematic Bipolar and Unipolar Switching Sequence. Direction of Rotation Viewed from Shaft End.

Compensation for this effect can be by either increasing the power supply voltage to maintain a constant current as the frequency increases, or by raising the power supply voltage and adding a series resistor as is shown below in the L/R drive circuit. Note that as the L/R is changed, more total power is used by the system.

The series resistors, R, are selected for the L/R ratio desired. For L/4R they are selected to be 3 times the motor winding resistance with a

watts rating = (current per winding)^2 x R.

The power supply voltage is increased to 4 times motor rated voltage so as to maintain rated current to the motor. The power supplied will thus be 4 times that of a L/R drive.

Note, the unipolar motor which has a higher coil resistance thus has a better L/R ratio than a bipolar motor.

To minimize power consumption various devices such as a bi-level power supply, or chopper may be used.

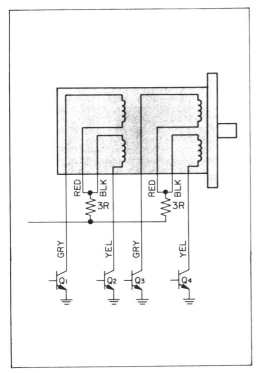

Fig. 15 L/4R Drive

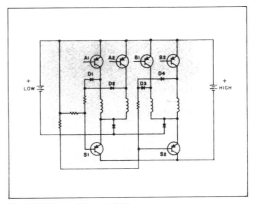

Fig. 16 Unipolar Bi-Level Drive.

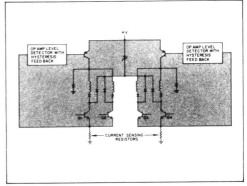

Fig. 17 Unipolar Chopper Drive.

BI-LEVEL DRIVE

The bi-level drive allows the motor at zero step/sec to hold at a lower than rated voltage, and when stepping to run at a higher than rated voltage. It is most efficient when operated at a fixed stepping rate. The high voltage may be switched on through the use of a current sensing resistor or by a circuit as shown below which uses the inductively generated turnoff current spikes to control the voltage.

At zero steps/sec the windings are energized with the low voltage. As the windings are switched according to the 4 step sequence, the suppression diodes D_1, D_2, D_3 and D_4 are used to turn on the high voltage supply transistors S_1 and S_2.

CHOPPER DRIVE

A chopper drive maintains an average current level through the use of a current sensor which turns on a high voltage supply until an upper current value is reached. It then turns off the voltage until a low level limit is sensed where it turns on again. A chopper is best for fast acceleration and variable frequency applications. It is more efficient than a constant current amplifier regulated supply. The V+ in the chopper shown in Fig. 17 typically would be five to ten times the motor voltage rating.

VOLTAGE SUPPRESSION

Whenever winding current is turned off, a high voltage inductive spike will be generated which could damage the drive circuit. The normal method used to suppress these spikes is to put a diode across each winding. This, however, will reduce the torque output of the motor unless the voltage across the switching transistors is allowed to build up to at least twice the supply voltage. The higher this voltage the faster the induced field and current will collapse, and thus the better performance. For this reason, a zener diode or series resistor is usually added as shown in Figure 18.

PERFORMANCE LIMITATIONS

Increasing the voltage to a stepper motor at standstill or low stepping rates will produce a proportionally higher torque until the magnetic flux paths within the motor saturate. As the motor nears saturation, it becomes less efficient and thus does not justify the additional power input.

The maximum speed a stepper motor can be driven is limited by hysteresis and eddy current losses. At some

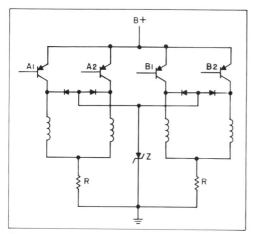

Fig. 18 Voltage Suppression Circuit.

rate, the heating effects of these losses limits any further effort to get more speed or torque output by driving the motor harder.

TORQUE MEASUREMENT
The output torque of a stepper motor and drive can best be measured by using a bridge type strain gage coupled to a magnetic particle brake load. A simple pulley and pull spring scale can also be used, but is difficult to read at low and high step rates.

MOTOR HEATING AND TEMPERATURE RISE
Operating continuous duty at rated voltage and current will give an approximate 40°C motor winding temperature rise. If the motor is mounted on a substantial heat sink, however, more power may be put into the windings. If it is desired to push the motor harder, a maximum motor winding temperature of 100°C should be the upper limit. Motor construction can be upgraded to allow for a winding temperature of 120°C (60°C rise).

SUMMARY OF KEY TORQUE EVALUATIONS

The speed-torque characteristic curves are the key to selecting the right motor and the control drive method for a specific application.

Define your application load.

Use the Start Without Error curve if the control circuit provides no acceleration and the load is frictional only.

Use the Running curve, in conjunction with a Torque = Inertia x Acceleration equation ($T = J\alpha$), when the load is inertial and/or acceleration control is provided.

When acceleration ramping control is provided, use the running curve and this torque equation.

$$T_J \text{ (Torque mNm)} = J_T \times \frac{\Delta V}{\Delta t} \times K$$

When no acceleration ramping control is provided, use the Running curve and this torque equation.

$$T \text{ (Torque mNm)} = J_T \times \frac{v^2}{2} \times K$$

Motor Selection Guidelines
1. **Based on frictional torque and speed make a first estimate motor selection.**
2. **Use torque equations and motion plot to evaluate.**
3. **If necessary, select another motor and/or modify the drive.**
4. **Secure prototype and test.**

Formula for determining temperature rise of a motor (using coil resistance) is:

Motor T rise °C =
$\frac{R \text{ hot}}{R \text{ cold}}$ (234.5 + T amb cold) - (234.5 + T amb hot)

Appendix E

Data Sheet

Miter and Bevel Gear Differentials

Differentials are simple planetary gear systems which inherently add or subtract angular movements transmitted to two of their components and deliver the answer to a third. They are widely used for adding and subtracting shaft movements in servo systems, and for addition and subtraction in computing machines. They can be geared with input and output shafts to multiply or divide inputs and outputs from and to these shafts. Differentials are also used to measure torque in a rotating shaft and to control the operation of other equipment.

Construction

Miter and bevel gear differentials consist of a "spider" and three bevel gears (Fig. 1). The spider comprises a shaft to which a cross-arm is rigidly fixed. One bevel gear is bearing-mounted on one end of the cross-arm. The other end of the arm carries a balancing block.

The other two bevel gears, bearing mounted on the spider shaft, mesh with the spider bevel gear.

The spider shaft is supported on shaft hangers and fitted at one end with a fixed spur gear for power transmission. The two bevel gears are also fitted with fixed spur gears for power transmission.

A miter gear differential has all three bevel gears of equal size and has a 1:1 ratio between the spider bevel gear and the other two gears. The bevel gear differential has a spider gear which differs in size from the other two, and therefore has other than a 1:1 ratio with them.

Spur gear differentials are also available but are less precise than the miter and bevel gear types due to unavoidable backlash. For example, in the case of the type shown (Fig. 2), teeth of the face gear are cut radially to the spider-shaft axis whereas those of the spider-arm spur gears are cut parallel to the spider-arm axis. Thus, there is not a perfect

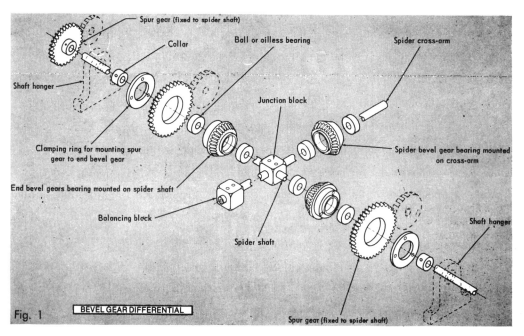

Spur gear (fixed to spider shaft)
Collar
Ball or oilless bearing
Spider cross-arm
Shaft hanger
Junction block
Clamping ring for mounting spur gear to end bevel gear
Spider bevel gear bearing mounted on cross-arm
End bevel gears bearing mounted on spider shaft
Balancing block
Spider shaft
Shaft hanger
Fig. 1
BEVEL GEAR DIFFERENTIAL
Spur gear (fixed to spider shaft)

mating of the meshing teeth and some play is unavoidable.

Operation

With reference to Fig. 3, assume that bevel gear A is held stationary, that the spider shaft is rotated clockwise and that bevel gear B is free to turn on its bearings. As the spider shaft rotates, the spider gear is rotated about axis YY. At the same time, it also rotates about the spider-arm axis XX. It will rotate about XX through an angle equal to the one through which the spider shaft is turned. Thus, we have two equal motions simultaneously carried to bevel gear B.

This gear therefore will turn through an angle twice the one through which the spider shaft was turned. (This will always be true regardless of the diameter of the spider gear.)

If we reverse the above situation, and rotate B with the spider free to turn, the spider shaft will rotate through half the angular displacement of gear B.

If bevel gear A is no longer held stationary but is rotated at the same time that B is rotated, it will affect the motion of the spider shaft and its angular displacement will equal one-half the vector sum of the angular displacement of the bevel gears.

$$D_S = \frac{D_A \leftrightarrow D_B}{2}$$

If both bevel gears rotate in the same direction at different speeds, the spider shaft will rotate in that direction at a speed halfway between the two and the differential adds.

If both bevel gears rotate in the same direction at the same speed, the spider gear will not rotate on the spider arm but the spider arm and shaft will rotate in the same direction as the gears and at the same speed.

MINIATURE DIFFERENTIAL

If the bevel gears are turned in opposite directions at different speeds, the spider shaft will turn in the direction of the more rapidly moving gear at one half the difference of the speed of the two bevel gears and the differential subtracts.

If the bevel gears are turned at the same speed

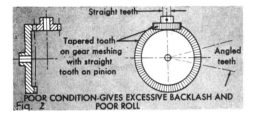

Straight teeth

Tapered tooth on gear meshing with straight tooth on pinion

Angled teeth

POOR CONDITION-GIVES EXCESSIVE BACKLASH AND POOR ROLL
Fig. 2

in opposite directions, the spider gear will turn but the spider arm and shaft will not.

The foregoing formula may, of course, also be used in determining outputs when the inputs are to the spider shaft and one bevel gear rather than to the two bevel gears.

In addition to the two input and output possibilities above, there is a third condition which can exist: to restrain the spider shaft and drive one of the bevel gears, the other bevel gear being

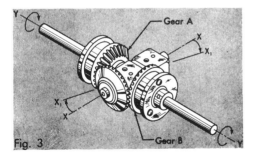

Gear A

X

X₁

X₁

X

Gear B

Fig. 3

free to turn on its bearings. Under these conditions the differential acts as a simple gear train which transmits motion from one bevel gear to the other but in the opposite direction. There will be a force tending to rotate the spider arm

equal to one-half the force transmitted from one bevel gear to the other, times the ratio of the spider bevel gear to the driving bevel gear. This force can be used to measure torque or to control equipment.

The differential can be used to multiply and divide by using differential spur gears of a different diameter than those with which they mesh.

For example, in Fig. 4, using the two bevel

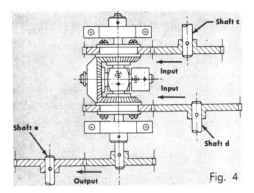

Shaft c

Input

Input

Shaft e

Shaft d

Output

Fig. 4

gears for inputs and the spider shaft for the output, if it is desired to obtain the algebraic sum of X times the rotation of shaft c plus Y times the rotation of shaft d, the ratio of the respective spur gears would be specified accordingly. By similarly specifying the ratio of the spider-shaft spur gear to its mate, the answer can be multiplied or divided. This can be expressed as follows:

$$D_s = \frac{XD_c \ +\!+\ YD_d}{2} \quad \text{and} \quad D_e = {}^\prime ZD_s$$

where D = the displacement of the respective shafts as indicated, and X, Y, and Z equal the ratios of the differential spur gears to the spur gears on shafts c, d, and e, respectively.

These formulas may also be transposed for use with other input-output combinations.

Applications

Of the many applications of miter and bevel gear differentials, perhaps the simplest is for changing the phase relation of one shaft with reference to others in the system. For example, in Fig. 5, if shaft A is geared through the differential as shown, the position of the output portion of the shaft with reference to the input section can be changed by means of a crank input. This action will alter the phase relation of the output section of shaft A with other shafts in the system. Phase changing can be done with everything fixed or with the shaft rotating.

Advancing one step further, miter and bevel gear differentials may also be used to alter the speed of a shaft. This is analogous to the phase shift application except that the crank input would be replaced by a continuously driven input. In this instance, the speed of the output section of the shaft differs with the speed of the input section, and the effective speed of the shaft with relation to others in the system is changed.

By relating the rotation of differential components to elements of computer equations, the bevel gear differential may be used to introduce

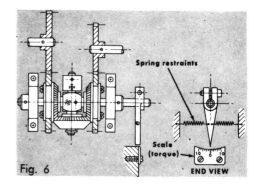

Fig. 6

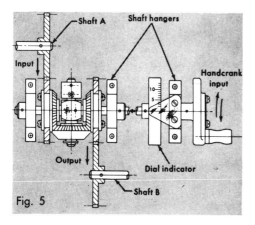

Fig. 5

data to computers without disrupting their operation. For example, referring to Fig. 5, computer information would normally flow in terms of shaft rotation into the differential via one bevel gear and out the other without alteration. If, however, it is desired to add or delete information, the hand crank is turned a distance which correlates with the data change to be incorporated, and the differential output is altered accordingly.

Miter and bevel gear differentials may also be inserted in a shaft system to measure torque, as shown in Fig. 6. The torque transmitted reacts against a calibrated restraint, thus providing a means of torque measurement.

Another application of miter and bevel gear differentials is as a clutch or brake. In this case the differential is inserted between the drive and the load, and one leg is restrained by means of

a clutch as shown in Fig. 7. As long as the load is less than a predetermined value, the clutch stays engaged and the load is driven. However, when the load exceeds the predetermined value, the transmitted torque opens the clutch and frees the restrained leg. This member, having no load, now rotates and the load remains stationary. Clutches can be arranged to engage and disengage as the load rises above and falls below a preset value.

Miter and bevel gear differentials are also used for error measurement and mechanical comparison by employing one leg to indicate differences in input to the other two.

They are also used in conjunction with pilot motors as speed controllers by arranging for the output leg to operate a rheostat (or valve) when the input speed from the prime mover deviates from the input speed from the pilot motor.

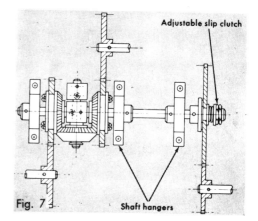

Fig. 7

Appendix F

Appendix F lists the product sources both by chapter and in alphabetical order for the convenience of the reader.

Chapter 1

Gaston County Dyeing Machine Company	Computer-controlled dyehouse Batch kiers and sequence dye machines MicroMonitor II Computer
Applied Color Systems, Inc.	INFOR/TEX

Chapter 2

Electronic Products Hearst Business Communications	Shielded data links and method of selection REPRINT
Belden Fiber Optic Division	Fiber optic cables and types of optical fibers

Chapter 3

Apple Computer, Inc.	Apple II + computer Apple interface cards
Rockwell International	PPS-4/1 single-chip MPU
Action Instruments Co., Inc.	Basic Controller BC-2

Chapter 4

Dilithium Press   Microcomputers for external
                                         control devices REPRINT

Datel, Inc.   Data acquisition and conversion
                                         data REPRINT

Chapter 5

National Semiconductor, Inc.   CMOS technology REPRINT

Chapter 6

Action Instruments Co., Inc.   Transpak
                                         Octapak
                                         Zilog MPU memory map
                                         Digital interface module
                                         Analog interface module
                                         BC-3 industrial computer

Haydon Switch and Instrument, Inc.   Big Inch stepper motor

Chapter 7

Micro Robotics Center, Inc.   Air-bearing slide positioner
Oriel Corporation   Stepper micrometers
                                         Dc-encoded motor micrometers
                                         Three-axis lens rotor
Octek, Inc.   Machine vision systems
Applied Intelligence Systems   PIXI-1000 machine vision
                                         system
Fairchild Camera and Instrument   CC-4001 machine vision camera
    Corporation   CCD image cell
Voran   VTR-6000 voice terminal
                                         VPC-2000 PC voice interface

Chapter 8

Cincinnati Milacron   Basic manipulator movement's:
                                         end effector yaw, pitch,
                                         and roll
                                         T3-566 and T3-586 robots
                                         Cinturn 12
Unimation Incorporated   Puma robot
                                         Apprentice robot
                                         Series 2000 heavy-duty robot

Chapter 9

American Photronics, Inc.   TR1000 fiber optic transmitter
                                          and receiver unit

Belden Fiber Optics Division   Bit-Driver modem
                                          Bit-Driver multiplexer
                                          Fiber optic cables

Burr-Brown Corporation   Fiber optic transmitters; 3712T,
                                          3713T, and FOT-110 KG-IR
                                          Fiber optic receivers; 3712R,
                                          3713R, and FOR-110 KG

Chapter 10

D.J.T. Electronics   Mini-73 single-board computer
H & R Corporation   Model robots: Line Tracer,
                                          Piper Mouse, and Space
                                          Invader

National Semiconductor Corp.   CIM-800 computer systems,
                                          Starplex and Starplex II
                                          development and operating
                                          language systems

RB Robot Corporation   RBX5 robot
Stock Model Parts   Memocon programmable robot
Technical Micro Systems, Inc.   BASYS/1 single-board computer
                                          ITSABOX computer-controlled
                                          robot

Terrapin, Inc.   Turtle robot
Wintek Corporation   Wintek 6800 modular computer

Chapter 11

National Semiconductor, Inc.   Digitalker model DT1050
                                          6502 CPU
                                          6522 VIA
                                          7475 flip-flop
                                          74150 data selector
                                          74154 data distributor
                                          LM339 comparator
                                          LM1812 sonar transceiver

Syntek, Inc.   SYM-1 Computer

Chapter 12

Scott Instruments Corporation   Shadow/Vet voice terminal

Veterans' Administration,   Quadriplegic wheelchair
Rehabilitation Engineering
Service

Sources

Action Instruments, Inc.
8601 Aero Drive
San Diego, CA 92123

Advanced Fiber Optics Corporation
636 South Hayden Road
Tempe, AZ 85281

Airpax Cheshire Division
Cheshire Industrial Park
Cheshire, CT 06410

American Photonics, Inc.
Milltown Office Park
Route 22
Brewster, NY 10509

Apple Computer, Inc.
10260 Bandly Drive
Cupertino, CA 95014

Applied Color Systems, Inc.
P.O. Box 5800
Princeton, NJ 08540

Applied Intelligence Systems
110 Parkland Plaza
Ann Arbor, MI 48103

Belden Fiber Optics Division
2000 South Batavia Avenue
Geneva, IL 60134

Winfred M. Berg, Inc.
499 Ocean Avenue
East Rockaway, NY 11518

Burr-Brown Corporation
P.O. Box 11400
Tucson, AZ 85734

Chartpak Graphics
One River Road
Leeds, MA 01053

Cincinnati Milacron
215 West Street
Lebanon, OH 45036

Datel, Inc.
1020 Turnpike Road
Canton, MA 02021

Dilithium Press
8285 South West Nimbus
Suite 151
Beaverton, OR 97005

D.J.T. Electronics
81 Orange Street
Sorrento, FL 32776

Electronic Products
Hearst Business Communications
645 Stewart Street
Garden City, NY 11530

Epson America, Inc.
3415 Kashiwa Street
Torrence, CA 90505

Lt. Cdr. H. R. Everett, U.S. Navy
Naval Sea Systems Command
Washington, DC 20632

Fairchild Camera and Instrument
Corporation
3440 Hillview Avenue
Palo Alto, CA 94304

Gaston County Dyeing Machine
  Company
P.O. Box 308
Stanley, NC 28164

H & R Corporation
401 East Erie Avenue
Philadelphia, PA 19134

Haydon Switch and Instrument, Inc.
1500 Meriden Road
Waterbury, CT 06705

Micro Robotics Center, Inc.
301 21st Street East
Bradenton, FL 33508

National Semiconductor Corporation
2900 Semiconductor Drive
Santa Clara, CA 95051

Octek, Inc.
7 Corporate Place
South Bedford Street
Burlington, MA 01803

Oriel Corporation
P.O. Box 872
Stratford, CT 06497

Robotics Age, Inc.
174 Concord Street
Peterborough, NH 03458

Scott Instruments Corporation
1111 Willow Springs Drive
Denton, TX 76205

Stock Drive Products
55 South Denton Avenue
New Hyde Park, NY 11040

Technical Micro Systems, Inc.
P.O. Box 7227
Ann Arbor, MI 48107

Unimation, Inc.
Shelter Rock Lane
Danbury, CT 06810

Veterans' Administration
Rehabilitation Engineering Center
252 Seventh Avenue
New York, NY 10001

Voran
4487 Technology Drive
Fremont, CA 94538

Answers to Odd-Numbered Review Questions

Chapter 1

1. Control of temperature by means of an electric timer driving a temperature-profile cam or template.

3. A device operated by compressed air; the air pressure to the device regulated by an electronic analog signal.

5. Analog signals vary constantly and over a wide range, whereas digital signals vary only in steps that may be considered either ON or OFF, like a switch.

7. The machine interface panel (MIP).

9. No. Load, sample, and unload remain manual functions.

11. Drain, safety interlock, and air pad.

13. To permit dye liquors to be heated above $100°C$ ($212°F$) without boiling or turning to steam.

15. They differ only in the control program. Both employ analog and digital control functions.

Chapter 2

1. Atmosphere, electromagnetic, electrical, and physical; electrical spike and hash interference.

3. They produce high-voltage spike pulses on the power lines when the motor is stopped and the magnetic fields surrounding the motor field winding collapse.

5. 1 to 3 microns thick.

7. Because of the lower dc resistance of braid over foil.

9. The distributive capacitance is lower in this method, which reduces transmission loss and enables longer cable lengths with the same efficiency of transmission.

11. **(1)** Elimination of electromagnetic and crosstalk interference.

    (2) No electrical ground loops.

    (3) No arcing.

    (4) Immune to electrical discharges.

    (5) Longer cable runs between repeaters.

    (6) No electrical hazards when cut.

13. Material dispersion.

15. The numerical aperture equals the sine of the phase angle, which is equal to the square root of the core index squared minus the cladding index squared.

17. Core center, 1.49; cladding interface index, 1.45.

19. Distortion produced by optical frequency variations producing propagation-time differences.

Chapter 3

1. The central processing unit, ROM and RAM, power source.

3. Typically, eight conductors in the form of a flat ribbon cable.

5. Parallel processing system.

7. 28- or 40-pin.

9. ROM and PROM or EPROM.

11. **(1)** Convert 2:00 P.M. to 140000 in 24-hour time.

    (2) Incorporate time in the IF . . . THEN statement as IF T = 140000 THEN TURNON RELAY 2.

    (3) DO UNTIL 150000 THEN TURNOFF RELAY 2.

13. The control program provides the host computer with the instructions for each slave computer then issues instructions to each slave unit in the time sequence specified by the control program. The slave computer contains specific control function instructions but requires start and stop commands from the host computer to execute the slave control functions.

15. Binary data where a logic 1 is +5 V dc and a logic 0 is 0 V dc or ground potential.

17. 10 change sense (3) TO = 1

    20 change sense (3) TO = 0

Chapter 4

1. Pressure, temperature, strain, or position.

3. By conversion of the analog signal to a digital representation of the analog data.

5. An 8-bit binary code representing any decimal number between 0 and 256.

7. Generally, a low-pass filter consists or either an *LC* or an *RC* network with the capability of transmitting or passing all frequencies below a cutoff point and blocking all frequencies above the cutoff point or frequency.

9. The four MSBs contain the decimal numbers from 16 to 256, whereas the four LSBs contain the decimal numbers from 0 to 15.

11. The A/D converter transforms an analog signal into its digital equivalent, whereas the D/A converter transforms a digital signal into its equivalent as an analog signal.

13. Successive approximation A/D converter.

15. The parallel conversion method is capable of a 25-MHz conversion rate for 4-bit data.

Chapter 5

1. PMOS.

3. The apparent increase in input capacitance by an amount equal to the voltage gain of the transistor.

5. Sodium contamination.

7. High speed and greater reliability.

9. Logic 0 input will produce logic 1 output (inverter).

11. Cost reduction, higher speeds, and better reliability.

13. Latch-up occurs when an input voltage is slightly higher than the power supply voltage or a negative voltage. In either case current will rapidly rise to a level that virtually shorts V_{cc} to ground and causes destruction of the IC.

15. Elimination of guard diffusions; reduction in propagation time, resulting in higher logic speeds and lower voltage requirements.

Chapter 6

1. The sensor would turn OFF the lamp when it turned ON because of the ambient room light level.

3. 3.5 to 5 V dc.

5. The voltage applied to the base, with respect to the emitter, that first begins to produce a current flow from the emitter to the collector. The point of forward bias is 0.03 V dc for germanium-based devices and 0.06 V dc for silicon-based devices.

7. The voltage drop produced by the relay coil's impedance.

9. Expand inputs to eight lines, transmit data signals to the computer, and condition the signal for transmission.

11. Platinum and platinum/rhodium.

13. One that runs continuously because of the high starting current drawn each time the motor starts.

15. The stepper motor rotates in fixed-degree steps, whereas the conventional motor runs at a continuous rotation rate.

17. 256 I/O ports.

19. Direct data bus connections and the I/O ports.

Chapter 7

1. Temperature coefficient of expansion.

3. Visible lines produced by the difference between a reference laser beam and a beam slightly out of phase in laser frequency because of a longer path from laser to subject.

5. The Oriel stepper-motor micrometer, because of the shock of operating in discrete angular steps.

7. Light-dependent resistors or photoresistive cells.

9. The smallest visible part of a video image.

11. 320 X 240 pixels.

13. By running a good part past the camera field several times to establish a reference signal data.

15. Increase the number of scanning cameras.

17. The maximum number of gray-scale levels is 16.

19. 325 ns.

21. An indication that the contents have been "dropped in."

23. A reverse image of the binary image.

25. Lower rejection rates, faster inspection, and inspection error as low as 0.001%, compared to human error of 12 to 15%.

27. Virtually a true reproduction of the user's voice quality in volume, tone, and speech characteristics.

Chapter 8

1. Resembling a person and the ability to move from one location to another under its own power.

3. Cartesian, cylinder, spherical, and articulated.

5. The shoulder center axis.

7. The "give" of an object when it makes hard contact with an immovable object.

9. Feedback data indicating its position or amount of movement.

11. High-pressure plumbing.

13. Five axes.

15. 1024.

17. To increase conveyor speed and subsequently increase production per shift.

19. Knowledge of the physical job to be done.

21. The T3-586.

23. By taking additional training to quality for new job skills or retrain for a specific job related to previous work.

Chapter 9

1. The heliograph.

3. Fiber optics.

5. Fiber optics, 8 dB-km; shielded coaxial cable, 125 dB-km.

7. A 14-pin rectangle design but differs in overall size.

9. PIN and APD.

11. 3.

13. 10 to 15 μW optical power.

15. Cooler operating of both transmitter and receiver.

17. No, because of the low attenuation during transmission.

19. +5 V dc and ground.

21. OFF.

Chapter 10

1. Electric motors, gear trains, leverage, and force.

3. Infrared optical energy.

5. Yes. A microprocessor is the only part of a computer where data may be entered, stored, or retrieved. To enter control functions into RAM requires a MPU.

7. Up to 35 ft.

9. The model electric train would require mobility without the aid of tracks, an on-board computer or umbilical cable attachment to a computer, and control facilities for maneuverability and collision/obstacle avoidance in order to be called a robot.

11. Using the minimum number of parts or control methods to accomplish the desired function.

13. For attachment to another device.

15. Making a section of the manipulator telescoping in action.

17. To change the line of action from one direction to another.

19. Modified 8th language.

21. Available compatible accessories and interfaces, software support, and accessible to high-level languages through control-compatible computer systems.

Chapter 11

1. Shakey used an SDS 940 time-sharing computer and not an on-board computer.

3. Self Contained Independent Mobile Robot.

5. Robart I (1981–1982).

7. SYN-1, 6502 MPU.

9. 280-word vocabulary.

11. Sixteen.

13. Direct data movement in or out of the VIA or PIA.

15. The Digitalker required 13 I/O lines.

17. The third control line is used with distributor A to overcome the nonlatching problem with distributor B.

19. Sixteen.

21. To turn the recharging station's beacon on and off.

23. Limit switches at each end of the scan direction; left or right.

25. Until one of the photocell comparators indicated that a light had been detected; indefinitely.

27. Reduced angle of view and improved signal-to-noise ratio.

29. Recharge electrical contact could be made from any approach direction.

31. To allow for momentary voltage drops by stalled motors.

33. The design eliminated the steering drive wheel in favor of dual drive wheels at midship and stabilizing swivel casters forward and at the rear of the wheelbase cage.

Chapter 12

1. 85% repeatability.

3. Voice entry may substitute for keyboard entry of data in response to the current program as long as the voice recognition vocabulary has been defined and is on the current operating disk.

5. 48k RAM.

7. By driver and driven sprocket gears coupled by a roller chain.

9. Navigation, voice terminal and computer accessibility, entertainment device control, environmental control, automatic battery-recharge function.

11. It must appear logical and should appear simple, repeatable, proportional, and reliable.

13. 97% error-free.

15. Faulty inspection practices, workmanship, and materials.

17. To support the total weight, approximately 20,000 lb.

19. Industrial diamonds or Carboloy teeth.

21. Steering is accomplished by increasing the speed of the tandem wheels opposite to the direction of steering.

23. Wet surfaces result in slippage and traction loss; suction-cup tread increases traction.

25. An insect's suction cup appendages on its feet.

27. Maintain automatic centering in passageways and hatches.

29. Painting of tall buildings, window washing, rescue, underwater search and mining, satellite recovery and repair, and all underground mining operations.

Index

A

Acquisition time, 93
Adjustments, to A/D and D/A
    converters, 108–10
Analog-to-digital (A/D) converters,
    100–6
  adjustments, 108–10
  counter or servo type, 100–101
  dual slope integrating type,
    101–3
  parallel type, 103–6
  successive approximation type,
    102
Analog multiplexers, 83, 106–7
  parameters, 107
Aperture time, 85
Arithmetic-logic unit, 57–58
Articulated design, robot, 184
Artificial intelligence, 175–77
  robot power sources, 189
  speech technology, 175–76
  voice output, 176–77

B

Basic controller BC2, 68–75
Baud rate, 67
Binary coded decimal, 96
Bipolar design, 125

C

Cartesian design robot, 184
Central processing unit (CPU), 55
Cincinnati Milacron Industrial Robot,
    195–202
  use in industry, 197
Common mode rejection ratio, 91
Complementary metal-oxide
    semiconductor (CMOS)
  technology, 116–19
    circuit performance, 118–20
    data converters, 125–27
    54/75HC logic family specifications,
      123–24

latchup problem solution, 121-22
output current sink and source
  comparison, 124-25
P2CMOS process, 120-21
speed improvement, 122-23
design, 125
in a logic circuit, 118-19
MDAC circuit, 127
Computer control
in a hostile environment, 26-51
  contaminants, 26-28
  fiber optics technology, 41-48
  peripheral data links, 29
  shielded data links, 30-40
Computer-controlled machine shop,
  19-23
Computer interfaces
for control and measuring devices,
  147-51
Computer peripherals, 52-79
  basic controller BC2, 68-75
  interface limitations, 54-62
  parallel and serial interfacing,
    63-68
  Zilog Industrial Basic Language
    (ZIBL), 75-78
Computer vision, 162-75
  requirements, 165-66
  video camera, 162-65
  video image capabilities, 166-75
Computer vision systems, 165-66
  cost, 172
  inspection, 173
Contaminants, 26-28
  classes of, 26-27
Control and measuring devices,
  135-44
  overshoot/undershoot characteristics,
    143-44
  temperature measuring devices,
    137-43
    mercury bulb thermometer, 138
    thermistor, 139

thermocouple, 138-39
thermostat, 139-43
Control devices, 130-47
  external devices, 132
  and measuring devices, 136-44
  operation requirements, 132-35
  transducers, 144-47
Cylindrical design, robot, 184

D

Data acquisition and conversion
  systems, 81-115
  amplifiers and filters, 89-92
  analog multiplexers, 106-7
  analog-to-digital and digital-to-analog
    adjustments, 108-10
  analog-to-digital converters,
    100-106
  digital coding, 94-96
  digital-to-analog converters, 96-100
  quantizing theory, 83-85
  sample and hold circuits, 107-8
  sampling theory, 86-89
  settling time, 92-94
design, of a teaching aid robot,
  229-37
  robot manipulators and end
    effectors, 234-37
  wheelbase designs, 230-34
Digital coding, 94-96
Digital-to-analog (D/A) converters,
  96-100
  adjustments, 108-10
Discrete I/O ports, 61
Dual slope A/D converter
  (*see* Integrating A/D converter)

E

External clock, 62

F

Fiber optic cables
  operating distances, 219–20
Fiber optic receivers, 215–17
Fiber optics technology, 41–48
  optical fibers, 43–48
    parameters, 45–48
  optical signal guidance, 41–43
  transmitters and receivers,
    207–10
    computer interfacing to,
    217–20
Fiber optic transmitters, 205–21
  data transmission and reception
    via optical frequencies, 206–7
  computer interfacing to, 217–20
  individual systems, 211–17
  and transceivers, 207–10

I

Industrial process, history of, 1–25
Industrial robots, 182–203
  of Cincinnati Milacron, 195–202
  components, 183–90
  on the production line, 191–95
INFOR/TEX, 12–15
  cost reduction, 12
  formula maintenance, 12
  inventory control, 12
  production control, 12
Instruction decode circuits, 62
Integrating analog-to-digital
    converters, 101
Internal clock (CLKIN), 62

L

Low-pass active filter, 83, 91

M

Machine computers
  installations, 151–53
M2CMOS process, 120–21
Mercury bulb thermometer, 138
Metal-oxide semiconductor (MOS),
    116
Micrometer, 158
Micromonitor II computer, 7–19
  control panels, 9–12
  drug room operator's panel, 11
  machine interface panel, 9
  machine operator's panel, 11
  main control panel, 9
  functions, 8
Micromonitor Process Control System,
    3–7
  computer, 7
  functions, 8
  control panels, 9–11
  INFOR/TEX software, 12–16
  process control analysis, 17–19
  program control operation, 17
Microprocessor (CPU) (*see* Central
    processing unit)
Mine rescue robot, 307
MPU memory map, 147–49

N

Nuclear power plant damage control
    robot, 301–6

O

Offset binary, 96
Optical fibers (*see also* Fiber optics)
  graded index, 43
  step index, 43

P

Parallel analog-to-digital converters, 103
Parallel and serial interfacing, 63–67
  Motorola 6280 PIA, 63–65
  peripheral interface adaptor (PIA), 63
  serial data transfer rate (baud rate),
    67
  serial RS-232-C interface, 65–67
P-channel metal-oxide semiconductor
    (PMOS), 116–19
Peripherals, computer, 53–79
  accumulator and arithmetic-logic unit,
    57–58
  central processing unit (CPU), 55
  control clock, 62
  conditional interrupts, 61
  data memory, 62
  discrete I/O ports, 61
  driver and receiver circuits, 58–59
  input channels, 61
  instruction decode, 62
  parallel processing systems (PPS),
    56–57
  program counter, 61
  read-only memory (ROM), 62
  SA register, 61–62
  S register and shift counter, 59
Peripheral data links, 29
Peripheral interface adaptor (PIA), 63
Program counter, 61
Programmer-sequencer circuit, 83
Project robots (*see also* Robot projects)
  with on-board computers, 237–43
    single-board computers for robots,
      243–46
P2CMOS process, 120–21

Q

Quantizing theory, 83–85

R

Random access memory, 62
Read-only memory, 62
Relay features, 71–75
Robart I
  design considerations, 258–83
    battery monitor circuit, 278
    battery-recharging capability, 270
    collision avoidance, 279
    data distributor operation, 265–66
    data selection operation, 263–65
    docking system, 276–78
    drive and steering control, 266–69
    head control and sensors, 269
    microprocessor selection, 259–60
    near-infrared proximity system,
      282–83
    PIA parallel interface layout, 261–63
    scan-and-track system, 271–76
    sonor collision avoidance system,
      280–82
    speech synthesis capabilities, 260
Robot projects, 254–89
  Robart I, 257–83
    design considerations, 258–59
  Robart II, 283–88
  SCIMR, 256
  Shakey, 254–55
  Stanford Cart, The, 255–56
  Unimation Rover, 256–57

S

Sample-and-hold circuit, 83, 107–8
  parameters, 108
Sampled data comparator circuit, 126
Sampling theory, 86–89
SA register, 61–62
S register, 59
Sensor inhibit control, 4–6

Settling time, 92-94
Shielded data links, 30-40
Shipboard damage control robot,
    307-12
Single-board computers for robots,
    243-46
    state-of-the-art, 246-50
Speaker-dependent recognition, 176
Speaker-independent recognition,
    175-76
Speech recognition, 175
Spherical design, robot, 184
Successive approximation A/D
    converter, 102

T

Teaching aid robot
    design, 229-37
        manipulators and end effectors,
        234-37
        wheelbase design, 230-34
    National Semiconductor Industrial
        Microcomputer System, 250-52
    project robots, 237-50
Temperature measuring devices, 137-43
    mercury bulb thermometer, 138
    thermistor, 139
    thermocouple, 138-39
    thermostat, 139-43
Test mode, 62
Thermistor, 139
Thermocouple, 138-39
Thermostat, 142-43
    overshoot/undershoot characteristics,
    143-44
Transducer
    definition, 144
Transmitter truth table, 214

U

Ultra-precision controlled devices,
    156-81
    artificial intelligence, 175-77
    light computer vision control,
        161
    machine and computer vision,
        162-75

V

Video camera, 162-65
Voice-entry terminal, 291-92
Votan voice terminal, 176-77

W

Wheelchair design projects, 292-301
Work robots, second-generation, 312

Y

Yaw, pitch, and roll movements,
    184-88

Z

Zilog Industrial Basic Language (ZIBL),
    75-77
    constants, 77
    functions, 76
    operators, 76
    statements, 75
    strings, 77
    special addresses, 77

DATE DUE